T0245102

Cambridge Elements

Elements in Emerging Theories and Technologies in Metamaterials

edited by
Tie Jun Cui
Southeast University, China
John B. Pendry
Imperial College London

INFORMATION METAMATERIALS

Tie Jun Cui
Southeast University, China
Shuo Liu
Southeast University, China

CAMBRIDGE
UNIVERSITY PRESS

University Printing House, Cambridge CB2 8BS, United Kingdom

One Liberty Plaza, 20th Floor, New York, NY 10006, USA

477 Williamstown Road, Port Melbourne, VIC 3207, Australia

314–321, 3rd Floor, Plot 3, Splendor Forum, Jasola District Centre,
New Delhi – 110025, India

79 Anson Road, #06–04/06, Singapore 079906

Cambridge University Press is part of the University of Cambridge.

It furthers the University's mission by disseminating knowledge in the pursuit of education, learning, and research at the highest international levels of excellence.

www.cambridge.org
Information on this title: www.cambridge.org/9781108958011
DOI: 10.1017/9781108955294

First published 2021

A catalogue record for this publication is available from the British Library.

ISBN 978-1-108-95801-1 Paperback
ISSN 2399-7486 (online)
ISSN 2514-3875 (print)

Information Metamaterials

Elements in Emerging Theories and Technologies in Metamaterials

DOI: 10.1017/9781108955294
First published online: January 2021

Tie Jun Cui
Southeast University, China

Shuo Liu
Southeast University, China

Author for correspondence: Tie Jun Cui, tjcui@seu.edu.cn

Abstract: Metamaterials have attracted enormous interest from members of the physics and engineering communities in the past 20 years, owing to their powerful ability in manipulating electromagnetic (EM) waves. However, the functionalities of traditional metamaterials are fixed at the time of fabrication. To control the EM waves dynamically, active components are introduced to the meta-atoms, yielding active metamaterials. Recently, a special kind of active metamaterials, digital coding and programmable metamaterials has been proposed that can achieve dynamically controllable functionalities using field programmable gate array (FPGA). Most importantly, the digital coding representations of metamaterials set up a bridge between the digital and physical worlds and allow metamaterials to process digital information directly, leading to information metamaterials. In this Element, we review the evolution of information metamaterials, mainly focusing on their basic concepts, design principles, fabrication techniques, experimental measurement and potential applications. Future developments of information metamaterials are also envisioned.

Keywords: metamaterial, digital coding, electromagnetism, information science, wireless communication

ISBNs: 9781108958011 (PB), 9781108955294 (OC)
ISSNs: 2399-7486 (online), 2514-3875 (print)

Contents

1 Introduction

1.1 Current Information Systems and Challenges

The history of information transmission can be traced back to 5,000 years ago when smokes and drums were used in China and Africa for long-distance communication. Such ancient information transmission approaches utilized human hearing and vision. Information transmission did not experience revolutionary progress until the invention of the telegram and the further discovery of electromagnetic (EM) waves in the nineteenth century, when people started to transmit signals through EM waves at the speed of light.

Figure 1.1 illustrates the architecture of the transmitter for the modern wireless communication system, which has been used for more than 40 years [1,2]. The information to be sent is first converted into digital codes, which can be easily stored and processed by electronic devices. Researchers and engineers in the wireless communications field have been focusing on how messages can be efficiently encoded and wirelessly transmitted to terminals at a maximum speed and with a minimum of errors. This goal is achieved by digital modulation, which is one of the most important modules in the communication system determining the transmission rate and the bit error rate.

Popular digital modulation techniques include amplitude-shift keying, frequency-shift keying, phase-shift keying [3] and quadrature amplitude modulation [4]. After the modulation, the digital baseband signals are represented by the variations in amplitude, frequency and phase of carrier waves. Because the frequency of the digital information is too low to be sent directly via a radio-frequency (RF) wave, it has to be converted to analog signals with a digital-to-analog converter (DAC) and then modulated to a high-frequency carrier wave via a mixer.

For brevity of content, some modules are omitted in the schematic shown in Figure 1.1, such as the digital up convertor, which upshifts the digitally modulated signal to digital intermediate frequency. The modulated signal containing the encoded information is further amplified by a series of power amplifier (PA) circuits and is finally radiated to free space through an antenna, as demonstrated in Figure 1.1. A receiver performs the reverse process of the transmitter.

In fact, all information-processing systems (e.g. wireless communication, radar and electronic confrontation) contain three subsystems: the baseband and digital signal processing (DSP) system; the microwave/RF system, an architecture of the transmitter for the modern wireless communication system; and the antenna, as presented in Figure 1.2. The first part is focused on the digital space belonging to the communication and signal-processing societies, while the second and third parts are focused on the physical space belonging to the

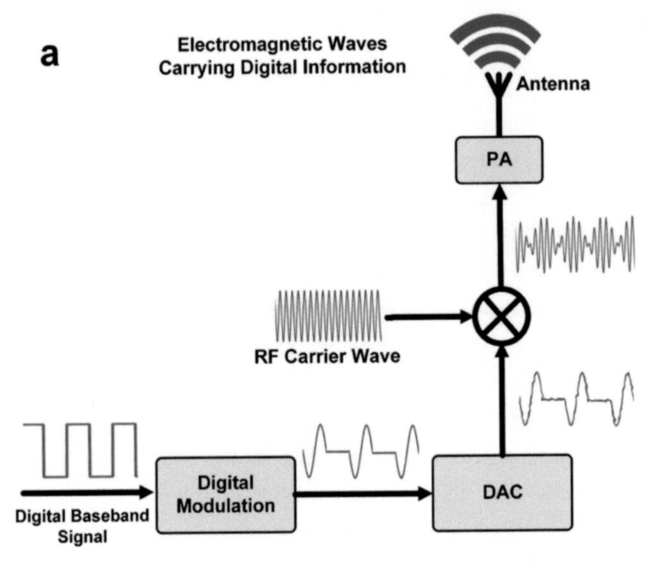

Figure 1.1 The architecture of the transmitter for the modern wireless communication system

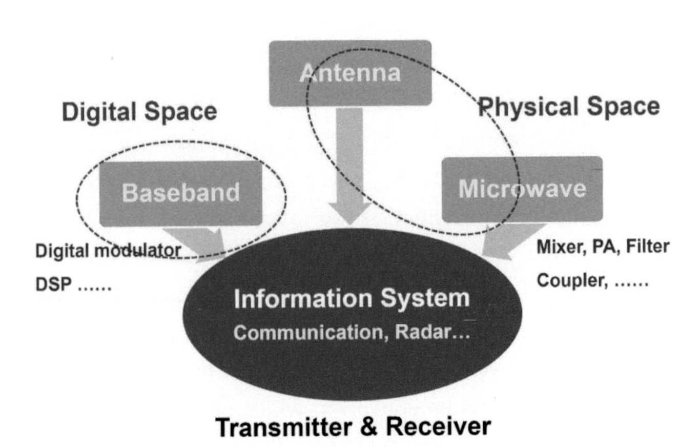

Figure 1.2 Basic modules required by all information systems

microwave and antenna societies. In the current electrical engineering and information technology communities (e.g. the Institute of Electrical and Electronics Engineers [IEEE]), the digital-world and physical-world societies are relatively independent, which causes limited system efficiency and considerable cost to the information systems. If the physical and digital worlds can be combined, we may expect higher system efficiency at lower system cost. Hence we should find an efficient way to bridge the physical and digital worlds, and metamaterial may provide an ideal medium.

1.2 A Brief Review of Metamaterials and Metasurfaces

Metamaterials are artificial structures composed of subwavelength unit cells in periodic or nonperiodic arrangements. The subwavelength nature of unit cells makes it possible to describe the metamaterials with effective medium parameters [5–7], which yields homogeneous effective media for periodic metamaterials or inhomogeneous effective media for nonperiodic metamaterials.

The flexible design of unit cells allows us to engineer medium parameters with unusual values, such as negative permittivity [8,9], negative permeability [10,11] and zero index of refraction [12–14], which do not exist in natural materials. Due to the flexibility of the spatial arrangement of artificial unit cells, it is possible to arbitrarily engineer the distributions of effective medium parameters, which is difficult or even impossible with traditional methods [15,16].

Two unique features of the effective medium parameters offer the metamaterials great potential in controlling EM waves or lights with unusual methods. With the emergence of transformation optics (TO) [17,18], a powerful theory concerning how to arbitrarily control the trajectory of light in the effective medium frame, the metamaterials have made possible many physical phenomena that had been only mentioned in science fiction in the past, such as perfect or super-resolution imaging [19,20], invisibility cloaking [21,22] and optical and EM illusions [23–25]. Many other exotic physical phenomena have also been widely realized and investigated, including negative refraction [26–28], negative reflection [29], EM diffusions [30], reversal Doppler effects [31] and EM black holes [23,33]. Metamaterials also have had a deep influence on the industry, bringing about many high-performance devices, such as high-gain antennas [34–38], flattened Luneberg lenses [39–41], gradient-index lenses [34,42,43], perfect EM absorbers [32,44–48], high-efficiency polarization converters [49–51] and reductions of radar cross-sections (RCSs) [52]. Hence metamaterials have provided a great impetus for research in both the science and engineering communities in the past 20 years.

The planar realization of metamaterials, commonly termed *metasurfaces*, is composed of arrays of optically thin elements periodically arranged on a planar surface. They could provide us with almost the same controls of the reflected and transmitted waves as the bulky metamaterials [53–60], and they possess many unusual EM properties due to the complex EM responses of their constitutive elements. Interestingly, such two-dimensional (2D) periodic structures can date back to the frequency selective surfaces (FSSs) [61], which are typically composed of dipoles or aperture arrays, and they were widely employed for the manipulation of EM waves from radio wave to optical spectra well before the emergence of metasurfaces. They have found widespread applications in microwave engineering, such as reflection/transmission array antennas [62], microwave filters and RCS reductions. However, FSSs did not receive wide attention until the emergence of metasurfaces, which are composed of elements at deeper subwavelength scale.

Metasurfaces are superior to metamaterials in the following aspects. First, it is difficult to integrate bulky metamaterials into compact devices due to their multilayered configuration, especially in the optical regime. Second, the fabrication of the three-dimensional (3D) nanostructures is still a challenge for current edging nano-fabrication technology such as electron beam lithography (EBL) and reaction ion etching (RIE), seriously inhibiting the development of bulk metamaterials at higher frequencies such as terahertz (THz), infrared and visible light. The thinly layered nature of metasurfaces makes them much easier for fabrication than the 3D metamaterials, and offers them potential integrations into the nano-photonic systems. Third, the wave manipulation of metamaterials relies on the phase accumulation of light propagation through a certain optical path, and inevitably suffers transmission loss due to the intrinsic lossy nature of metals and dielectrics near the resonances. Such losses lead to strong absorptions of EM waves, which deteriorate the device performance – or even make it impossible – for some exciting applications such as perfect lensing [63,64] and transformation optical cloaks [17,18,21,22]. The metasurfaces, however, are found to have much lower losses, and hence show better performance in manipulating the EM waves [55–57].

Although the loss of metasurfaces should be minimized to reach higher efficiency, especially for the manipulation of transmitted waves, they can also be designed as lossy for developing perfect EM absorbers. The past decade has seen great interest in researching and designing various metasurface absorbers working from the microwave [44], through THz [65] and infrared [66–67], to the visible light [68–71] spectra. The metasurface absorbers can operate in flexible ways, such as in single-band [44,65], dual-band [72,73], multiple-band [74–76] and broadband [45,77,78] modes, demonstrating potential applications in bolometer, photodetector and photovoltaic devices.

Wavefront shaping is probably the most inspiring functionality of the metasurfaces. By spatially arranging specially designed meta-atoms on the planar surface, we can achieve a complete 2π phase shift for either the co-polarized or cross-polarized component of the incident beam. In 2011, Yu et al. rewrote the conventional laws of wave refraction and reflection (i.e. Snell's law) by considering the phase discontinuities introduced on the interface between two media using V-shaped antennas with different opening angles [54]. The deliberately designed V-shaped antenna is characterized by two geometrical parameters, the length of rods and the angle between them, which provide enough degrees of freedom for tailoring the amplitude and phase of the cross-polarized wave to realize the anomalous reflection and refraction.

The V-shaped antenna was later employed to design a broadband, background-free quarter-wave plate, which could convert the arbitrarily linearly polarized incident beam into high-quality circularly polarization and reflect it to anomalous direction [79]. More details of the V-shape antenna and its analytical model can be found in Refs. 54,58,60,79–83. Important applications of the wavefront-shaping metasurfaces include a vortex beam generator [84], holographic imaging [85–90] and all-dielectric magnetic mirror metasurface [91], to mention just a few.

Recently, Capasso's group achieved highly efficient focusing of visible light into diffraction limited spots using an ultrathin metasurface-based optical lens called the metalens [92]. It is fabricated on a glass substrate with an array of TiO_2 nano-fins, which has a thickness of only 1/100,000 of conventional lenses. This metalens featuring ultrathin thickness and excellent performance was selected as one of the top 10 breakthroughs in 2016 by *Science* magazine, and it has great potential to replace conventional optical lenses in the near future [93–95]. Note that although the concept of a generalized Snell's law is new, the physics behind it can be actually classified into the category of Pancharatnam-Berry (PB) phase, which was discovered earlier in 2001 by Hasman's group for achieving space-variant manipulation of the light polarization state using computer-generated space-variant subwavelength metal and dielectric gratings [96,97].

The other benefit of the PB phase is that it offers us a 2π phase control of the cross-polarized transmission waves, while this cannot be realized by any single-layered nonmagnetic metasurfaces, which can only provide a phase coverage from 0 to π. For the PB phase, it is known from the Poincare sphere that there would be an additional phase difference in the range $0\sim\pi$ when the polarization state of light changes from the linear horizontal polarization to linear vertical polarization, or from the left circular polarization to right circular polarization, in which the transmitted wave is orthogonally polarized with respect to the incident wave [97]. Although the PB phase can be used to realize the 2π phase coverage for the cross-polarized transmitted wave, the maximum attainable efficiency is theoretically limited by 25% due to the fact that there is only one electric surface. In optics, the efficiency normally reaches only 10% due to the intrinsic losses in the plasmonic antenna structures. To improve the transmission efficiency for the manipulation of co-polarized component, Monticone et al. suggested a symmetric stack of three metallic layers that has access to the full 2π range for the control of co-polarized transmitted waves [98].

A new approach, the Huygens surface, can fundamentally improve the transmission efficiency of the co-polarized component to nearly 100% by introducing the interference of electric and magnetic polarizability on the surface using a dual-layered configuration [99]. This idea was inspired by the surface equivalence principle by Schelkunoff in 1936 [100], who stated that if there are two regions, each having a different wave propagation and separated by an infinitesimally thin interface, the discontinuities of the field component on the boundary can be compensated by a fictional surface with certain electric and magnetic currents. However, fabricating a metasurface with given electric and magnetic properties is a challenging task, especially in the THz and optical frequencies, because the plane of structures has to be perpendicular to the magnetic field of the incident wave, which does not comply with the conventional photolithography technique.

Another important application of metasurfaces is polarization manipulation, such as the conversion from linear polarization to circular polarization [71], or from horizontally linear polarization to vertically linear polarization [49]. In addition to various optical devices, the metasurface polarizers have wide applications in dual-polarized radars, fiber-optic communications, wireless communications and millimeter wave imaging systems, as they provide an additional information channel for radio and optical systems.

The ultrathin thickness and flexibility of the metasurfaces have also inspired Alù and coauthors to design a new cloaking technique called a mantle cloak [101–103], which is a metasurface warping around a dielectric cylinder to cancel the scattering from the inner dielectric cylinder. Despite its much thinner thickness compared to the TO-based cloak, it can produce almost the same cloaking performance. This technique was later extended to cloak the conducting cylinder with tunable cloaking frequencies [104,105]. The metasurfaces can also be used to design a hologram, a technique widely used to record and reconstruct the 3D wavefront of an object by storing and releasing its phase/amplitude information.

Larouche et al. [106] demonstrated the metasurface phase hologram composed of arrays of pixels with different effective index in the infrared frequency, in which the light passing through them gets the designed phase shift, thus forming the corresponding holographic image in the image plane. Recently, there have been some reports on the metasurface-based broadband 3D holographic imaging [88,107–109], and single holographic metasurface incorporating multiplexed images [86,90,110–112].

In addition to the metasurfaces with fixed geometrical parameters of unit cells, it is very desirable to control EM properties dynamically. The unit cell is typically required to have tunable phase and/or amplitude responses under some external controls. Several approaches have been demonstrated for implementing the tunable metasurfaces, including photo excitation of free carriers, electro excitation of free carriers, phase change materials, mechanical movement and liquid crystal (LC).

Examples of tunable metasurface designs are included in what follows. Minovich et al. proposed controlling the optical properties of fishnet metasurfaces using nematic LC [113], in which the reorientation of LC molecules inside the fishnet holes leads to the frequency shift of the hole mode. Another LC-based tunable metasurface was demonstrated by Decker et al. [114] for featuring 90° polarization rotation. By controlling the external electric field across the LC cell, one could switch EM modes of the unit cell, which further results in a change of the transmission coefficient. The EM properties of some materials are sensitive to temperature, such as VO_2, which exhibits dielectric properties under room temperature but undergoes a change to conducting state at ~67°C. Such a unique feature was utilized to develop a thermally tunable metasurface [115], which shows a dramatic change in the reflection spectra when the sample is gradually heated. The phase tunability of the unit cell can also be realized by changing the physical shape of the unit cell structure. Feasible methods include the mechanical system [116], micro-electromechanical systems (MEMS) [117–122] and microfluidics system [123,124].

Compared with the mechanically tunable metasurfaces, one may be more interested in the electrically tunable metasurfaces without moving parts, which have much faster response speeds and are more stable. In the microwave frequency, such electrically tunable metasurfaces can be realized by loading active components to the unit cells, such as pin-diodes and varactors [125–127]. At higher frequencies like THz or infrared, however, the deep subwavelength discrete elements are not applicable. In these cases, semiconductors [128], ferrites [129], graphene [130–133] and phase-changing materials [134] are the best substitutes for the electrically tunable metasurfaces.

For example, n-type gallium arsenide (GaAs) substrate and gold electric split-ring resonators (SRRs) were employed to create a Schottky diode [128], enabling the metasurface to function as a THz modulator with transmission up to 50%. Chen et al. experimentally realized amplitude modulation of the THz waves by doping a thin layer of p-type GaAs on an intrinsic GaAs substrate [135]. This work was later extended to a spatial light modulator that has 4×4 individual pixels, each independently controlled by an applied voltage [136].

1.3 Digital Coding and Programmable Metamaterials

As mentioned in the previous paragraphs, the metamaterials and metasurfaces characterized by the effective medium parameters have powerful capabilities in controlling the EM waves due to the arbitrary medium parameters and their spatial distributions. However, they cannot control the EM waves in real time. In other words, once the metamaterials and metasurfaces are fabricated, their functionality will be fixed for the passive case, or they will be tunable only with similar functions for the active case. To achieve the real-time manipulations of EM waves and produce many but completely different functions with a single metamaterial or metasurface, one should consider a new way to characterize the metamaterials.

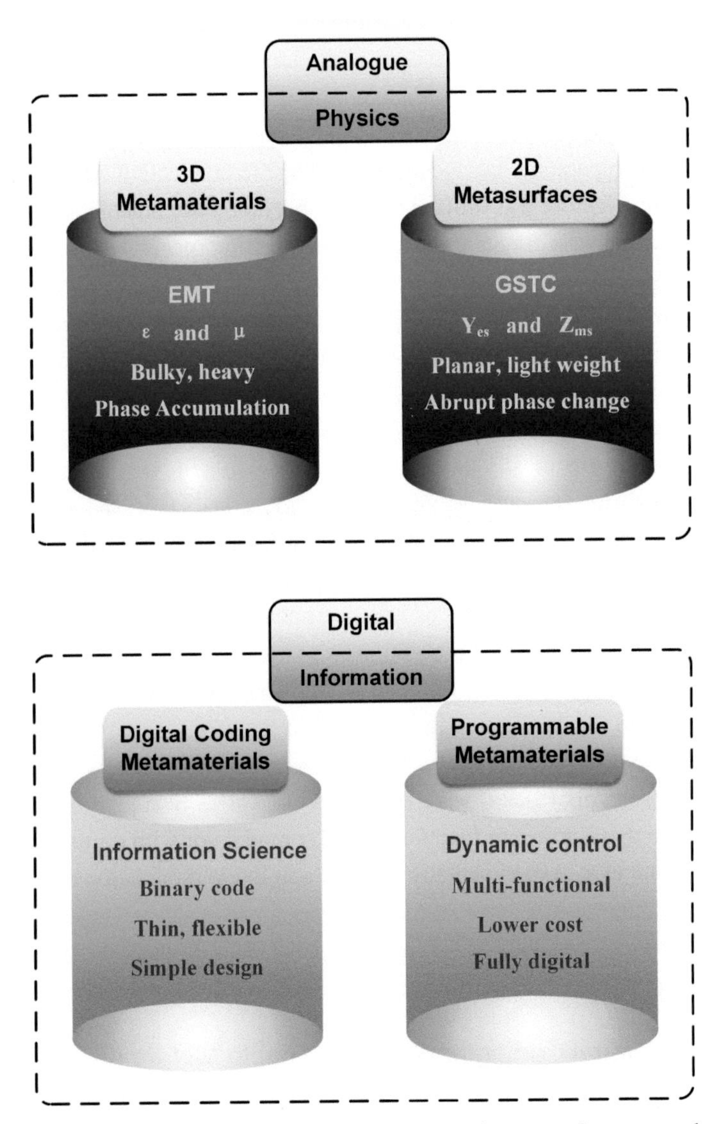

Figure 1.3 The development of metamaterials and metasurfaces over the past 20 years

Before we introduce the concept of digitally coding metamaterials and metasurfaces, it is necessary to summarize the metamaterials and metasurfaces by looking back on their development over the past 20 years, as sketched in Figure 1.3. Metamaterials, due to their 3D configurations, can manipulate light through the phase variation along the optical path. The fabrication difficulty and inevitable loss of the 3D metamaterials hinder their development, especially in the higher frequency range, thus pushing the emergence of thinner and lighter versions of metamaterials – metasurfaces. The way metasurfaces manipulate light has experienced a dramatic change from the gradual phase accumulation to the abrupt phase change, with performance comparable with or, in some cases, even superior to that of the metamaterials.

If we further investigate their characterization techniques, we notice that the metamaterials are mainly characterized by the effective permittivity and permeability retrieved from the effective medium theory [137,138]. However, those attempts to model metasurfaces as bulk media using the effective medium theory have proved to be unsuccessful due to their ultrathin thickness. Later the metasurface was suggested to be described by susceptibility, which is the average value of the surface polarizations proportional to the macroscopic (tangentially averaged) electric and magnetic fields crossing the thin metasurface [139]. The unique susceptibility is almost independent of the incident angle, and can be retrieved from the generalized sheet transition conditions (GSTCs) [139].

Since both the effective medium theory and GSTC method derive from physical principles that require sophisticated analysis and are difficult to combine with the existing digital systems, it would be interesting to study the metamaterials and metasurfaces from a digital perspective. From the circuit point of view, the effective medium parameters can be compared to analog circuits because they have continuous values. As we know, the analog circuits have obvious drawbacks such as worse noise-resistant performance and lower signal precision. Most importantly, the analog circuits require sophisticated theoretical design and sufficient engineering experience to achieve certain functions. Digital circuits, on the contrary, can process and deliver signals with much higher precision because they take only two, or a discrete number 2^n levels of a signal. Engineers no longer need to get into the physical details of each logic circuit unit, but only need to pay more attention to how to realize various functionalities with a series of simple logic operations.

To realize the digital version of the metamaterials and metasurfaces, we proposed the concept of digital coding and programmable metamaterials and metasurfaces in 2014 [140], in which the constituent blocks – i.e. digital coding particles – are described in a more direct manner with quantized reflection or transmission phases. Similar to the digital circuit, the reflection or transmission phases of the digital coding particles are defined as 0° and 180° for the 1-bit coding metasurface, and the phase is equally divided into 2^n levels across the 2π range if we want to build the n-bit coding metasurface. These digital coding particles are arranged on a 2D surface according to a given coding sequence, where the collective responses of the individual properties of each coding particle give ultimate control over beam propagation and polarization. In other words, the desired distributions of the EM fields in both near and far regions can be achieved through certain digital coding sequences.

The digital representation of the reflection phases makes coding metamaterials compatible with digital logic devices such as pin-diodes. By loading one or multiple pin-diodes to the coding particle and switching their states to ON and OFF, we can dynamically control the reflection or transmission phases. When an array of the digital particles is independently controlled by a field programmable gate array (FPGA), we can have a real-time manipulation of the impinging wave with desired functionalities simply determined by the input digital coding sequences. This is the programmable metamaterials that were initially proposed and experimentally demonstrated in the microwave frequency in 2014 [140].

Someone may doubt the similarities between the programmable metamaterials and the conventional phased array antennas [141,142] regarding their functionalities. However, the programmable metamaterials and phased array antennas are significantly different. The element of phase arrays typically falls in the order of 1/2~1 wavelengths or larger while for the digital coding metamaterials, it is usually four times smaller, around 1/8~1/4 wavelengths. The smaller unit cell of the programmable metamaterials allows a wider scanning angle of the radiated beam, and, most importantly, supports conversion from spatially propagating wave to surface wave.

The most distinguishing characteristic of the programmable metamaterials is the fully digital description, which profoundly changes the way we understand, analyze, and design metamaterials. Once the structure design of the digital particle is finished, engineers only need to design different coding sequences to manipulate near-field distributions and far-field radiations. This means that the same coding sequence adapts to different types of coding particles, different operational frequencies, and even different kinds of waves. This characteristic quite resembles digital circuit design, where only the code design of hardware description languages (VHDL, Verilog), but not the physical realization of the logic gate, should be concerned. It is worth mentioning that the digital description of the coding metamaterials allows us to revisit metamaterials from the perspective of information science. Owing to the Fourier relation between the coding pattern and the radiation pattern, many theorems and analytical tools in digital signal processing and information science can be directly applied to the analysis and design of the digital coding metamaterials, enabling more exotic physical phenomena and interesting functionalities.

1.4 Information Metamaterials

The twenty-first century is the era of information technology, which is one of the fastest-growing fields in all of the scientific disciplines and has a deep influence in almost every aspect of our daily life. The combination of digital coding metamaterials and signal processing algorithms gives us access to new research directions from the information science perspective and to many unconventional devices and systems. To further combine the digital coding and programmable metamaterials with information science, we proposed the concept of information metamaterial and metasurface in 2017 [143,144]. This is a kind of metamaterial or metasurface that can process the digital coding information directly under the illumination of EM waves, and can further achieve the transmission, processing, and self-learning of digital information.

As illustrated in Figure 1.4, the digital coding information to be transmitted is directly mapped to the digital coding pattern on the information metasurface, which is actually converted to the control signals of the digital metasurface. Under the incidence of EM waves, such digital coding information is modulated to the scattered EM waves and transmitted to the far field by the information metasurface. In this manner, we can build up a digital world (digital coding information) directly on the physical world (metasurface). Hence the wave-metasurface interaction could also be the wave-digital

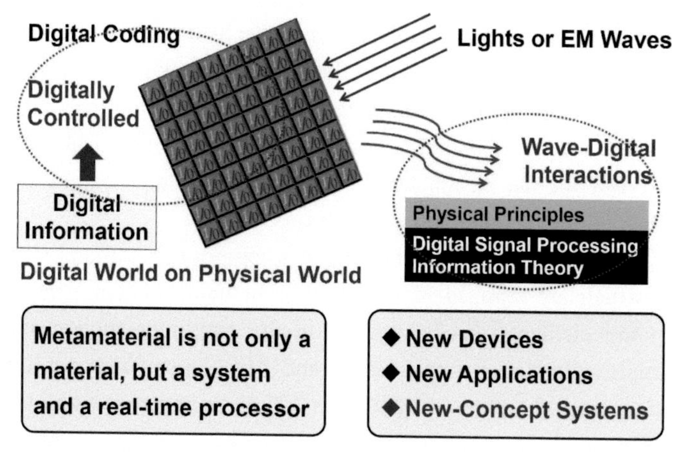

Figure 1.4 Working principle of the information metamaterials

interaction, ensuring that both physical principles (for the physical world) and digital signal processing methods (for the digital world) can be used to manipulate the EM waves and digital information. This will result in new devices, new applications, and even new information systems. Therefore, the information metamaterial is not only an effective material but also a system with real-time information processing ability, which sets up a bridge between the physical world and the digital world, providing an alternative solution to the current information systems.

1.5 Outline of This Element

In this Element, we present various types of information metamaterials reported in the past six years for wide operational frequencies ranging from microwaves to THz, and for different wave types from EM waves to acoustic waves, focusing primarily on the basic concepts, important functions, design methods, fabrications, and experiment validations. We also examine the recent progress of the information metamaterials in imaging and communication systems with new system architectures. In the last chapter, we envision some of the intriguing implications of the information metamaterials on different fields from radio waves to optics, and discuss some potential techniques that are suitable for implementing the programmable metamaterials. We conclude with an outlook on directions for future developments of the information metamaterials.

2 Digital Coding Metamaterials

2.1 Digital Coding Representation of Metamaterial: Basic Concept

Traditionally, metamaterials are described by the effective medium parameters due to their subwavelength unit cells, and they can be considered as continuous media in the macroscope. As stated in Section 1.3, this kind of metamaterials can be regarded as “analog” metamaterials. To deliver “digital” version of metamaterials, we have to find a new characterization approach. Here, we propose to describe the metamaterial structure using digital coding elements to reach the digital metamaterial.

As shown in Figure 2.1a, the metamaterial is composed of “0” and “1” coding elements, which are realized in many different manners. For example, the digital coding elements can be defined in the amplitude modes, with “0” element representing zero transmission and “1” element representing total transmission; or in the phase modes, with “0” element representing 0° phase and “1” element representing 180° phase, etc. Then we can control the EM responses, in both the near field and the far field, of the metamaterial by using

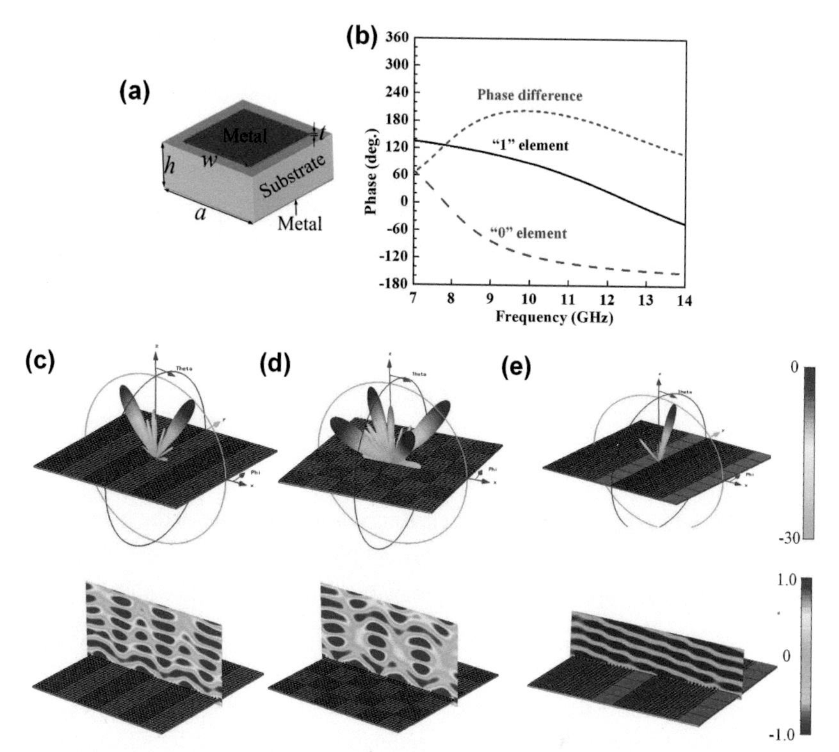

Figure 2.1 Microwave coding metamaterials. (a) Structure of 1-bit coding metamaterial. (b) Reflection phases of the "0" and "1" element from 7 to 14 GHz. (c, d, e) Far-field radiation patterns of the "010101 . . . ", "010101 . . . /101010 . . . " (chessboard), and "00 01 10 11 00 01 10 11 . . . " coding sequences, respectively. (f, g, h) The near-electric-field distributions correspond to the radiation patterns in (c), (d), (e), respectively.

different digital coding sequences of "0" and "1," instead of specific medium parameters. Since "0" and "1" coding elements have only two binary states, we name the corresponding metamaterial as a 1-bit digital coding metamaterial.

This concept can be easily extended to multi-bit digital coding. For instance, 2-bit digital coding metamaterial is composed of four coding elements: "00," "01," "10" and "11," which have phase responses of 0°, 90°, 180° and 270°, respectively; while 3-bit digital coding metamaterial is composed of eight coding elements: "000," "001," "010," "011," "100," "101," "110" and "111," which have phase responses of 0°, 45°, 90°, 135°, 180°, 225°, 270° and 315°, respectively. For the general case of an n-bit digital coding metamaterial, it is composed of 2^n coding elements, and the phase difference between adjacent coding elements will be $360°/2^n$.

2.2 Microwave Digital Coding Metamaterials

Although the first coding metamaterial was experimentally validated in the microwave frequency, we emphasize that the analogous concepts and techniques can be extended to higher frequencies or applied to other types of waves, for example, the THz regime[145–147] and acoustic wave [148–150]. In this section, we start with the first coding metamaterial, which is implemented at microwave frequency, describing its basic concept, working mechanism and design strategy. Then, different physical realizations of coding metamaterials for THz and acoustic waves are introduced in the subsequent sections, together with some new physical phenomena and interesting functionalities.

2.2.1 Design of Digital Coding Particles

Figure 2.1a shows the structure of 1-bit coding particle designed at the microwave frequency, which is composed of three layers: the bottom metallic ground sheet, the middle dielectric spacer and the top metallic square patch. The metal ground sheet helps prevent the incoming wave from interacting with the materials behind it, and helps improve the amplitude of reflection at the same time. The reflection phase can be arbitrarily tuned in a wide range when the side length of square patch varies. For the 1-bit case, two different coding particles with different lengths of square patch are required. They are carefully designed to reflect the impinging wave with 0° and 180° phases, and are designated as "0" and "1" elements.

Figure 2.1b provides the reflection phase spectra of the "0" and "1" elements in a broad band from 7 to 14 GHz, where we can see that the phase difference reaches exactly 180° at 8.7 and 11.5 GHz, and fluctuates between 135° and 200° from 8.1 to 12.7 GHz. Note that only the phase difference between the "0" and "1" elements, but not their absolute phases, affects the performance of coding metamaterials. Of course it is very desirable to develop coding particles with broadband performance, for example, with constant 180° phase difference in a broad bandwidth for the 1-bit case. However, passive structures support very limited bandwidth for the 180° phase difference due to their intrinsic dispersion. Non-foster components could be potentially applied to expand the working frequency band of coding metamaterials by compensating the reactance of the passive structure [151–153].

Because the 0 and 1 coding particles fully reflect incident waves in phase (0°) and out of phase (180°), respectively, if we view the digital coding metasurface as an impedance sheet, then the 0 and 1 coding particles can be viewed as a perfect electric conductor (PEC) and a perfect metallic conductor (PMC), respectively. Note that a coding metasurface with all 0 coding particles mimics the high impedance surface (HIS), which is usually composed of corrugated metallic structure with quarter-wavelength deep grooves [154] or mushroom-type structures over a ground plane [155], and could help antenna maintain high radiation efficiency when it is placed close to the reflective surface.

2.2.2 Control of EM Waves Using Digital Coding Sequences

In what follows, we show how we manipulate the EM wave through various arrangements of these coding particles (i.e. coding sequences or coding patterns) on the 2D plane. Two representative 1-bit coding sequences and their radiation patterns are given in Figure 2.1. The first coding sequence is "010101 . . . ", which deflects the normally incident wave in two symmetrical directions, as can be observed in Figure 2.1c. The second coding sequence "010101 . . . /101010 . . . ", called the chessboard coding pattern, splits the normal incidence into four beams pointing in four symmetrical directions (see Figure 2.1d). The radiation patterns of these coding patterns, which are composed of $M\times M$ equal-sized lattices with dimension D, can be analytically calculated using the following function,

$$Dir(\theta,\varphi) = \frac{4\pi|f(\theta,\varphi)|^2}{\int_0^{2\pi}\int_0^{\pi/2} |f(\theta,\varphi)|^2 \sin\theta d\theta d\varphi} \tag{2.1}$$

$$f(\theta,\varphi) = f_e(\theta,\varphi)\sum_{n=1}^{N} \exp\{-i\{\varphi(m,n) + kD\sin\theta[(m-1/2)\cos\varphi + (n-1/2)\sin\varphi]\}\} \tag{2.2}$$

in which the scattering function of the deep-subwavelength coding particle becomes vague in the far field, and thus can be neglected in Equation (2.2). Note that each lattice (i.e. each binary code) in the coding pattern, which is also termed a *super unit cell*, is not a single coding particle, but contains a sub-array of $N\times N$ identical coding particles.

The use of super unit cells benefits the design of coding metamaterials. On one hand, as the phase response of a coding particle highly relies on the EM interaction with its neighboring ones, the real phase response will deviate from the designed value if its neighboring coding particles have different geometrical parameters. Super unit cells consisting of an array of identical coding particles can establish the local periodicity condition to mimic the infinite periodic structure, which is close to the unit cell boundary condition settings in the CST Microwave Studio for the simulation of a single unit cell. The larger the super unit cell, the more accurate the phase response of each coding particle. On the other hand, we can easily control the radiation direction by changing the size of the super unit cell according to the principle of scattering pattern shift [156], discussed in Section 5.3. The size of the super unit cell can be utilized to control the level of diffusion of the random scatterings for the controllable random surface (see Section 5.4) [157], or the opening angle of the cone-shaped radiation (see Section 5.5) [158].

For the 1-bit coding metamaterial, the radiation pattern is always symmetrical with respect to the normal axis. To generate a single-beam radiation pointing in an arbitrary direction, at least 2-bit coding particles are required, which have four different phase responses: 0°, 90°, 180° and 270°. They can be readily obtained by finding the corresponding lengths of square patch in Figure 2.1a. Apparently, one can design an n-bit coding metamaterial by finding a total number of 2^n geometrical parameters with $360°/2^n$ phase interval. However, for practical applications, 2-bit or 3-bit coding metamaterial is sufficient to generate various radiation patterns, as has been demonstrated by the many simulations and experiments presented in the following sections. Although we consider the reflection phases for the first example of coding metamaterial, we remark that the transmission phases can also be discretized to make transmission-type coding metamaterials, as introduced in the next section.

Figure 2.1e shows the radiation pattern of the gradient coding sequence "00 01 10 11 00 01 10 11 ... ", where the normally incident beam is redirected to the anomalous direction. The angle of anomalous reflection is determined by the generalized Snell's law [110], and can be arbitrarily designed according to the principle of scattering pattern shift [156]. The aforementioned case can be viewed as a one-dimensional (1D) coding sequence in which the code varies along a single direction and the single-beam can only scan in the elevation angle. Apparently, for a 2D gradient coding sequence, in which the code varies along both the x and y directions, the beam can scan in both the elevation and azimuthal planes.

These examples belong to the periodic coding patterns, which contain several repetitions of a certain coding sequence. They can redirect the incident beam to one or multiple definite directions. Some applications may require a low RCS detection, which can be simply realized with a random coding pattern. Figure 2.2a shows the fabricated sample of a 1-bit random coding pattern in which the coding sequences along the horizontal and vertical directions are the same "00110101" sequence. The simulated radiation patterns are shown in Figures 2.2b and 2.2c for 8 and 11.5 GHz, respectively. We can see that the incident wave is diffused into multiple beams pointing in random directions. The broadband diffusion performance of the random coding pattern is further demonstrated in Figure 2.2d, which shows the simulated and measured radar cross-section (RCS) reduction (with respect to a bare perfect electric conductor (PEC) board of the same size) in the backscattering direction from 7.5 to 13 GHz. From the simulation result it is clear that the 1-bit random coding pattern can reduce the backscattering RCS by at least

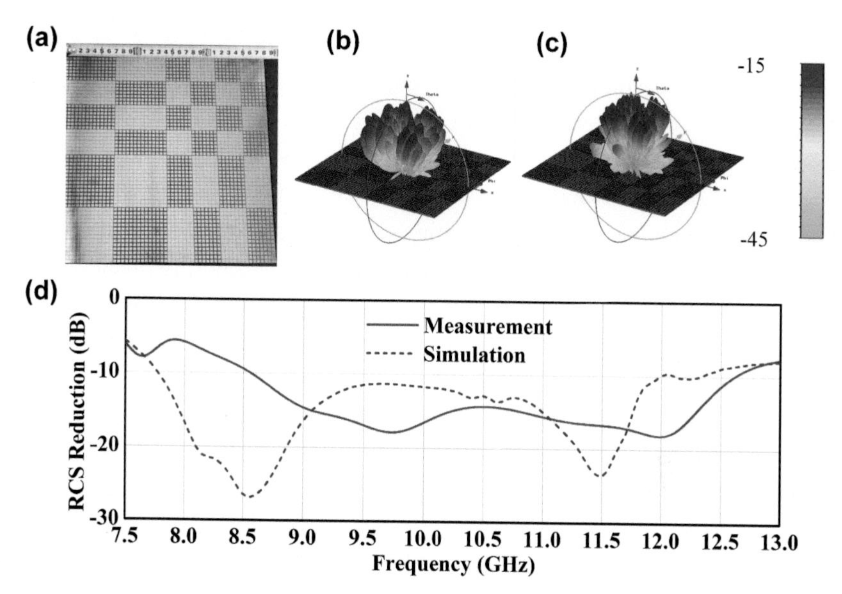

Figure 2.2 Random diffusion effect of the coding metamaterial. (a) Fabricated sample of the microwave coding metamaterial with 1-bit random coding pattern. (b, c) The simulated radiation patterns at 8 and 11.5 GHz, respectively. (d) The simulated and measured RCS reduction in the backscattering direction from 7.5 to 13 GHz.

10 dB in broadband from 7.8 to 12 GHz, which roughly coincides with the working bandwidth of the single unit cell. Note that in the initial work of coding metamaterial, we only consider the RCS reduction in the 0° direction. Some applications may require a low RCS in the specular direction, as discussed in Section 2.12.

Of course it would be more beneficial if the RCS could be suppressed in all possible directions, making the target low detective under the bi-static radar system. By employing Golay-Rudin-Shapiro (GRS) polynomials, Moccia et al. found a finely optimized coding pattern that uniformly distributes the energy in the upper-half space to realize a low level of scattering in all visible angles [159]. Taking advantage of its planar configuration, low cost and ultrathin nature, we believe this technique is more competitive in practical applications than previous techniques such as TO-based cloaking [17,18,21,23] and scattering cancellation–based cloaking techniques [101,102,160,161].

2.3 Terahertz Digital Coding Metamaterials

In this section, we consider two different coding metamaterials that are designed and fabricated at the THz frequency. As we discussed the basic concept and working mechanism of coding metamaterial in the previous section, we focus primarily in this section on the physical realization and experimental characterization of THz coding metamaterials.

Although we can directly upscale all the microwave designs to the THz regime by scaling down the structure in Figure 2.1a, it is difficult to find appropriate materials for the dielectric spacer. Most commercial dielectric substrates are designed for microwave and can hardly support frequency over 500 GHz, as the dielectric loss becomes remarkable at higher frequencies and seriously reduces efficiency. Researchers have found some low-loss dielectric materials at THz frequency, such as intrinsic silicon, intrinsic GaAs, sapphire, quartz, etc. Because the wafer thickness of these materials is typically in the range of 300–500 μm, they are highly rigid in nature and are commonly used as the carrier to provide mechanical strength for undertaking the micro-fabrication processes. But as they belong to the crystalline material, it is challenging to reduce their thickness down to 80 μm, which is still too thick for the THz coding metamaterials to work properly at a frequency of 1 THz or higher.

It is difficult to fabricate a dielectric spacer with only several microns, or even tens of microns, using current state-of-the-art micro-fabrication techniques. Grinding the 500 μm-thick wafer to 30 um thick is definitely impossible, while growing such materials to the thickness of 30 um using chemical vapor deposition (CVD) technique can be very expensive and impractical. Polyimide can serve as an appropriate material for the dielectric spacer of coding metamaterials at THz frequency, which can be easily fabricated with thicknesses of 2~40 μm at low cost using a spin-coating and baking process. Owing to the excellent flexibility of polyimide, it can bend, twist, or even be conformal with curved objects. Moreover, it has very stable chemical properties as well as superior mechanical stretchability. Such excellent features make the polyimide coding metamaterials very promising for applications of biomedical detection and wearable electronic devices.

2.3.1 Minkowski-Loop Digital Coding Metasurface

The first THz coding metamaterial was designed to make broadband diffusion of THz waves using a first-order Minkowski loop (see Figure 2.3a), which features wider operation bandwidth than the square patch design. By changing the geometrical parameters of the Minkowski closed loop, reflection phases of 0°, 45°, 90°, 135°, 180°, 225°, 270° and 315° can be obtained to build 1-bit, 2-bit or 3-bit coding metamaterials, as illustrated in Figures 2.3b and 2.3c. All the line width and gaps of the Minkowski closed-loop geometry are larger than 5 μm so as to be compatible with the currently available micro-fabrication

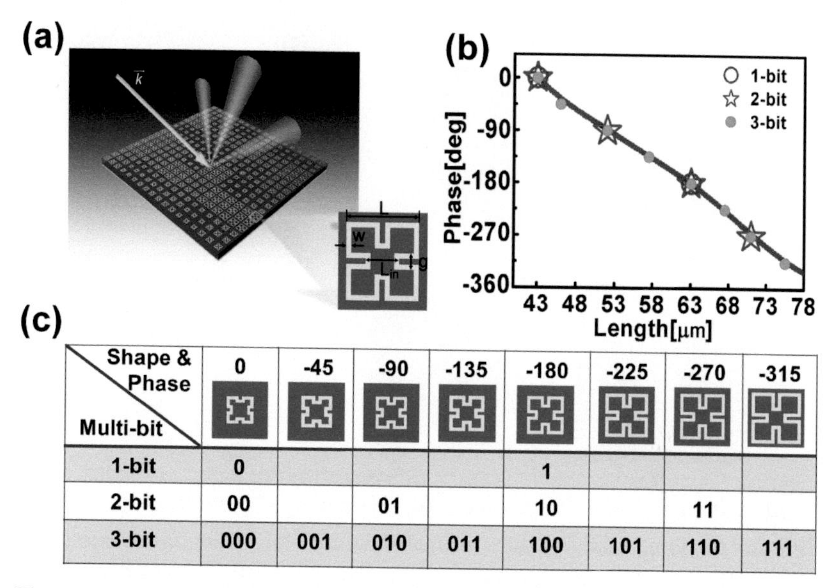

Shape & Phase / Multi-bit	0	-45	-90	-135	-180	-225	-270	-315
1-bit	0				1			
2-bit	00		01		10		11	
3-bit	000	001	010	011	100	101	110	111

Figure 2.3 The first THz coding metamaterial. (a) Structure of the THz coding metamaterial using Minkowski loop structures. (b) Reflection phases of the 1-bit, 2-bit and 3-bit coding particles as a function of length L. (c) The optimized structures for the 1-bit, 2-bit and 3-bit THz coding particle.

techniques. The thickness of the metal is set as 200 nm so that we can neglect the skin depth effect. In fact, no observable differences are found between the experimental performance of the samples fabricated with 200 nm-thick and 60 nm-thick metallic layers.

To demonstrate the diffusion performance of the THz coding metamaterial, two different random coding patterns generated with 1-bit and 3-bit coding particles are shown in Figures 2.4a and 2.4d, respectively. Both coding patterns have the same size of 7.56×7.56 mm^2, equivalent to an electrical size of $25.2\lambda\times25.2\lambda$ at 1.0 THz. Figures 2.4b and 2.4c show the 3D and 2D radiation patterns of the 1-bit coding pattern at 1.1 THz, in which the multiple scattering beams demonstrate the perfect diffusion behavior of the THz coding metamaterial. Because the 2D simulated radiation pattern (see Figure 2.4b, which is extracted from the E-plane of the 3D radiation pattern) has been normalized to the total reflection reference (fully deposited with Au layer), a 20 dB RCS reduction can be observed at the 0° direction. Similar performance can also be obtained by the 3-bit random coding pattern (see Figure 2.4d) at 1.1 THz, with the simulated 3D and 2D radiation patterns given in Figures 2.4e and 2.4f, respectively.

Next, we briefly introduce the process for the sample fabrication of THz coding metamaterial, which is schematically illustrated in Figure 2.5a. First, 10 nm-thick

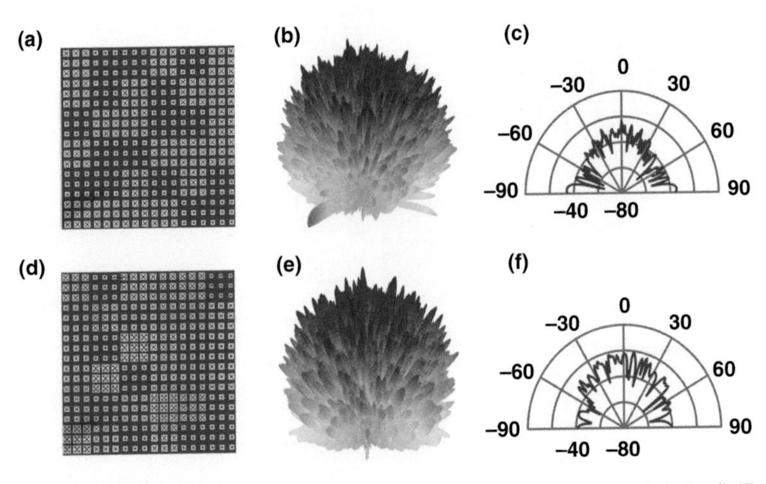

Figure 2.4 Diffusion of THz wave with THz coding metamaterial. (a,d) Two different random coding patterns using 1-bit and 3-bit coding metamaterials, respectively. (b,e) 3D radiation patterns of the random coding patterns in (a) and (d), respectively. (c,f) 2D simulated radiation pattern extracted from the E-plane of the 3-D radiation patterns in (c) and (f), respectively.

titanium (Ti) and 200 nm gold (Au) are deposited onto a silicon wafer using E-beam evaporation to serve as the metallic ground plane. Then, liquid polyimide is spin-coated onto the metal ground, by which a curing process on a hotplate at 80°C, 120°C, 180°C and 250°C (each temperature for 5 minutes) is followed to solidify the polyimide layer. Because the maximum thickness of the polyimide layer that can be made in a single spin-coating process is about 10 μm, a total number of three of the same spin-coating processes is needed to form the 30 μm-thick polyimide layer. Next, LOR-10A and AZ-5214 photoresists are spin-coated onto the wafer, followed by soft-bake, exposure, and development processes to form the required photoresist pattern. Then, another 10/200 nm Ti/Au layer is deposited on the photoresist pattern. In the final stage of the process, the sample is immersed in acetone liquid, with an ultrasonic bath of about 1 minute to help form the final metallic pattern. The final sample is shown in Figure 2.5b.

Broadband diffusion performance is experimentally demonstrated by measuring the reflection spectra of the fabricated sample with a custom-built time-domain (TD)-THz system. The schematic of the THz system is illustrated in Figure 2.5c, which primarily consists of a pair of fiber-coupled photoconductive antennas, one for emission and another for detection. The laser source generates an 84 fs laser pulse at the central wavelength of 1,550 nm. The laser pulse, after it is coupled into a dispersion-compensating fiber, splits into two beams, the pump beam and the probe beam. The movement of the gradient-index (GRIN) lens coupler produces the

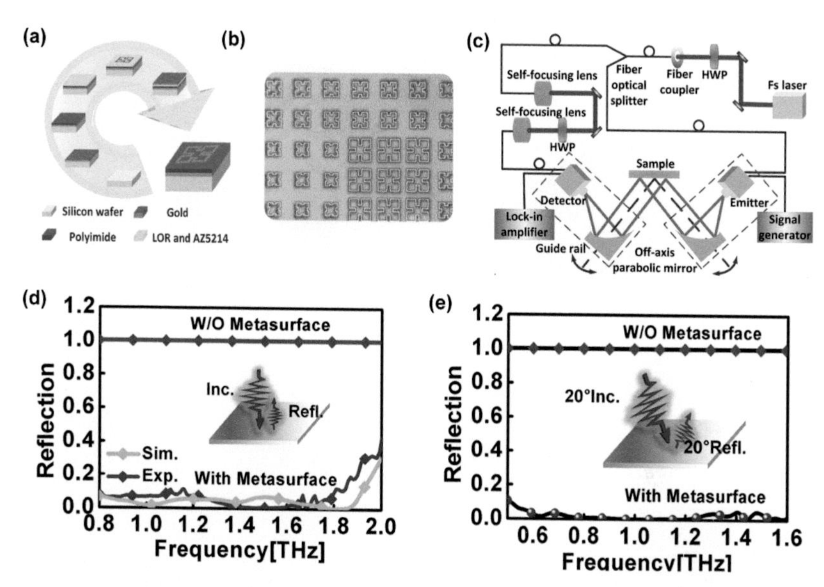

Figure 2.5 Fabrication and experimental characterization of THz coding metamaterials. (a) Fabrication process of the THz coding metamaterials. (b) Microscopy image of the fabricated sample. (c) Schematic of the custom-built time-domain THz system. (d, e) The measured reflection spectra of the sample with 2-bit random coding pattern at the normal (0°) and 20° oblique incidences, respectively.

required temporal delay between the THz pulse and probe beam, which is critical for the TDS-THz system to obtain the TD-THz signal. Two off-axis parabolic mirrors, mounted on guide rails, can collect and collimate the THz beams to the sample at any oblique incident angle. During the experiment, the transmitting and receiving antennas can be freely rotated in the horizontal plane to measure the reflection spectra of the sample at different angles. Figures 2.5d and 2.5e show the measured reflection spectra of the sample encoded with a 2-bit random coding pattern at the normal and 20° oblique incidences, respectively. It is clear that the reflection in both cases keeps lower than 10% of the specular reflection of the reference case in a broad bandwidth from 0.8 to 1.8 THz. The excellent broadband diffusion performance of the THz coding metamaterial may be desirable for the application of RCS reduction.

2.3.2 Round-Loop Digital Coding Metasurface

The other work on the broadband diffusion of THz wave was reported by Liang et al. in 2015. Instead of controlling the geometrical parameters of coding particle, they proposed to build the "1" and "0" elements by designing a unit

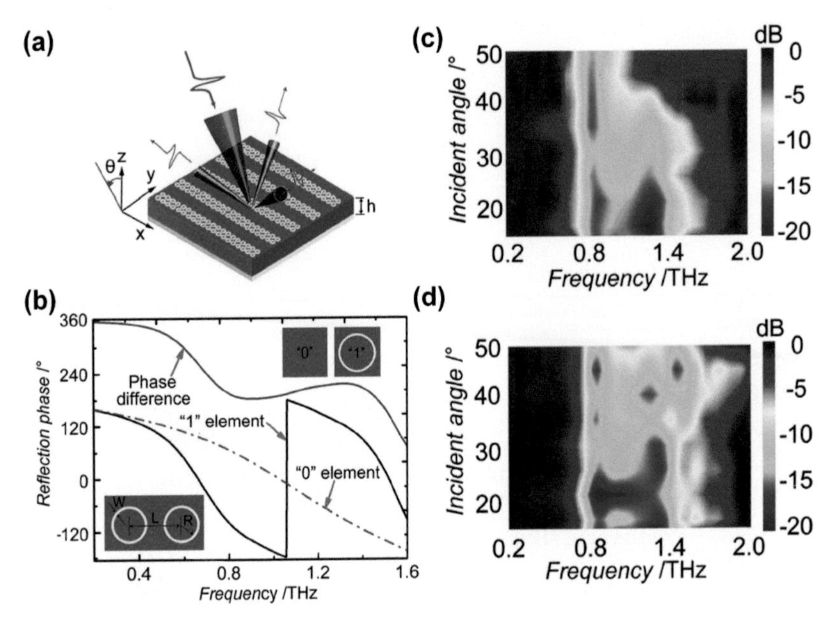

Figure 2.6 Terahertz coding metamaterial based on ring resonators. (a) Schematic illustration of the THz coding metamaterial in making random scattering of THz waves. (b) Structure and reflection phases of the "0" and "1" elements. (c) The measured reflection spectra of the flat sample when the incident angle varies from 13° to 50°. (d) The measured reflection spectra of the curved sample when the incident angle varies from 13° to 50°.

cell with and without a metallic ring resonator, as illustrated in Figures 2.6a and 2.6b, respectively. Similar to the unit cell design in the previous subsection, the 1-bit coding particle is made of a sandwiched structure with a metallic ring resonator, a 40 μm-thick polyimide layer, and a 200 nm-thick gold film, from the top to the bottom. It can be observed from the reflection curves in Figure 2.6b that the phase difference between the "1" and "0" elements maintains around 180° in a wide frequency band from 0.8 to 1.4 THz, which helps the coding metamaterial keep a stable functionality in a broad bandwidth.

To experimentally validate the performance of the coding metamaterial in making beam splitting and random diffusion with periodic and random coding patterns, the authors fabricated the samples with the similar process given earlier. A similar THz-TD system was employed to experimentally characterize the broadband performance of the fabricated sample. Figure 2.6c shows the measured reflection spectra of the flat sample under a wide oblique incident angle range from 13° to 50° with transverse electric (TE) polarization. The bandwidth of RCS reduction keeps almost stable for incident angles smaller than 40°. At 13° oblique incidence, the specular reflection is reduced by at least

−10 dB in the relatively broad bandwidth. Similar performance was also achieved for the transverse magnetic (TM) polarization case.

Another novelty of this work is the flexibility of the sample, which can be wrapped onto curved objects. Since RCS reduction for curved surfaces is more commonly encountered than flat surfaces in real applications, the authors considered the conformal case in their work by wrapping the sample around a metallic cylinder with a diameter of 37 mm. Figure 2.6d shows the measured reflection spectra of the conformal sample under TE polarization as the incident angle increases from 13° to 50°. Significant RCS reduction is observed in a wide frequency band for incidence angles up to 45°. In addition to the challenges in fabricating flexible coding metamaterial with a certain curvature, the accurate modeling and simulation of curved coding metamaterial is another obstacle that currently hinders the research of conformal coding metamaterial, due to the enormous number of mesh grids produced by the finite-difference–time-domain (FDTD) method. A fast calculation method with limited computation time and computational complexity should be developed to evaluate the performance of curved coding metamaterial.

2.4 Acoustic Digital Coding Metamaterials

Due to the similarity between EM waves and acoustic waves, and the rapid development of metamaterials in the past decades, there has been increasing interest in the manipulation of acoustic waves using acoustic metamaterials, including soft 3D acoustic metamaterials with negative index [1], 3D broadband acoustic ground cloak [2] and nonreciprocal and highly nonlinear active acoustic metamaterials [3]. The recent advances and future directions in the field of acoustic metamaterials can be found in Ref. 4.

2.4.1 Design of Acoustic Digital Coding Metasurface

In 2016, Xie et al. presented the first design and experimental demonstration of an acoustic digital coding metasurface [149], which could split or focus acoustic waves in a certain bandwidth, as illustrated in Figure 2.7a. Figure 2.7b shows the fabricated sample of the "0" (yellow) and "1" (blue) elements of the acoustic coding metamaterial. Both elements share a similar structure, which consists of two solid plates separated by a small gap, and six rectangular grooves etched on the upper plate. Two narrow slits are cut at the lower and upper plates, which serve, respectively, as the entrance and exit for the acoustic waves that are incident from underneath. The similar interior structure of both elements guarantees a nearly identical transmission amplitude for both elements, while the different position of the exit slits of the "0" and "1" elements determines the

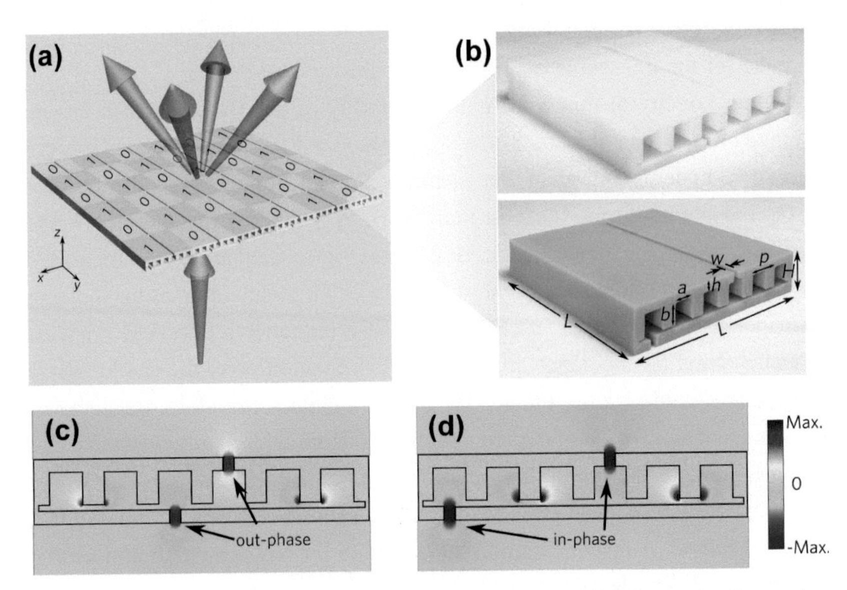

Figure 2.7 The first acoustic coding metamaterial. (a,b) Schematic illustration and structure of the acoustic coding metamaterial, respectively. (c,d) The velocity field distributions of the "0" and "1" elements at the resonant frequency.

180° phase difference between them. The authors claim that since only two distinct elements are involved in the 1-bit acoustic coding metamaterial, it is easier to keep the opposite phase difference between both elements in a relatively broad bandwidth than it is in conventional metasurfaces that have multiple elements.

The structure is simulated with the finite-element method in commercial software, COMSOL Multiphysics. All the materials are treated as acoustically rigid, with sound speed $c_0 = 343$ ms^{-1} and mass density $\rho_0 = 1.2$ kgm^{-3} set in the simulation. Numerical results show that the amplitude decreases sharply away from the resonance, but the π phase difference could maintain in a relatively broad bandwidth. The transmission peaks in the spectra rely on the excitation of local modes within the space between the nearest bumps inside the structures. This is different from conventional acoustic metasurfaces based on Fabry-Perot resonance.

Figures 2.7c and 2.7d give the velocity field distributions of the "0" and "1" elements at the resonant frequency, respectively. For the "0" element, it can be observed that the wave at the exit slit is out of phase with the wave at the entrance slit (see Figure 2.7c), while it is in phase for the "1" element (see Figure 2.7d). The strong horizontal mass vibrations in the narrow gaps between

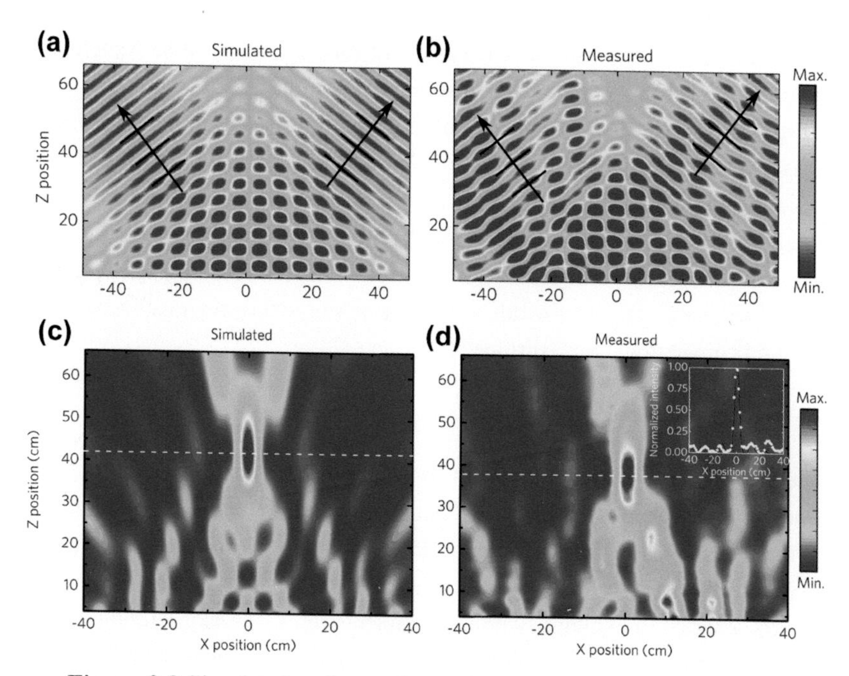

Figure 2.8 Simulated and experimental results of the acoustic coding metamaterials in refracting and focusing acoustic waves. (a,b) The simulated and experimentally measured near-field distributions of the sequence "101010 . . . " at 4,500 Hz. (c,d) The numerically simulated and experimentally measured field at 4,530 Hz of flat Fresnel zone plate.

the upper plate and the bumps of the lower plate indicate the occurrence of resonance. The π phase retardation is guaranteed by the length p, which corresponds to the point $k_x p/2\pi$ in the first dispersion band. One of the advantages of this design is that the "0" and "1" elements share a similar structure, making it possible to realize acoustic programmable metamaterials by opening or shutting down the slit mechanically.

The experiment was conducted by placing a 1D coding metasurface in a waveguide and illuminating it with a Gaussian beam. Figures 2.8a and 2.8b show the simulated and experimentally measured near-field distributions of a sequence "101010 . . . " at 4,500 Hz, which were obtained by scanning an rectangular region of 104×66 cm at a height of 2 cm. Similar to the microwave coding metamaterials, the normally incident acoustic wave is scattered into two branches pointing in symmetrical directions with respect to the normal axis, as indicated by the black arrows. The excellent agreement between the simulations and experiments validates the effectiveness of the design. The authors also present a flat Fresnel zone plate based on the acoustic coding metamaterial. The optical path difference between each adjacent zone should be half a wavelength to produce constructive interference at the focal point.

Figures 2.8c and 2.8d show the numerically simulated and experimentally measured fields at 4,530 Hz. It can be clearly observed from both simulation and experiment that the normally incident wave is focused to the focal point with a full width at half maximum (FWHM) of 4.6 cm. Owing to the broadband nature of the element design, the authors stated that the focusing effect can be maintained in a certain bandwidth, which is superior to the previous Fresnel lens based on conventional metasurface [162].

2.4.2 Acoustic Asymmetric Transmission Device

Based on the similar structure design for the acoustic coding metamaterial, the same group designed a multiband asymmetric transmission device by changing the arrangement of elements "0" and "1" [150], as shown in Figures 2.9a and 2.9b, respectively. Although topological acoustic metamaterials have been recently reported to exhibit asymmetric transmission by breaking time-reversal symmetry [163–166], their thickness is comparable to the wavelength. In this work, the total thickness of the acoustic coding metamaterial is only about $\lambda/6.3$. In the numerical simulation, the plate is considered a perfectly rigid body due to the large impendence mismatch with respect to air. For simplicity, the corrugated side is defined as the positive incidence (PI) and the smooth side is defined as the negative incidence (NI). The degree of transmission asymmetry R_c is defined as,

$$R_c = \frac{T_{PI} - T_{NI}}{T_{PI} + T_{NI}} \tag{2.3}$$

where T_{PI} and T_{NI} represent the transmission for PI and NI, respectively. Figures 2.9c and 2.9d show the amplitude and phase of the coding sequences "0000 . . . " and "1111 . . . " for both PI and NI. No asymmetric transmission occurs in either of these cases because they have exactly the same transmission, resulting in a zero value of R_c. Figure 2.9d shows a π phase difference between the "0000 . . . " and "1111 . . . " sequences. As the coding sequence becomes "0101 . . . ", the phenomenon of asymmetric transmission appears at three different frequency bands around 4,508, 5,204 and 5,274 Hz, as can be observed from Figure 2.9e. The first asymmetric band originates from the lower-order resonance, whereas the second and third asymmetric bands result from the higher-order effect. High-contrast ratio (from 0.7 to 0.8) is achieved at the three asymmetric bands.

To confirm the performance with an experiment, a sample composed of seven "1/0" units was fabricated by the fused-deposition-modeling 3D-printing technique, and placed in a planar waveguide system surrounded by acoustical

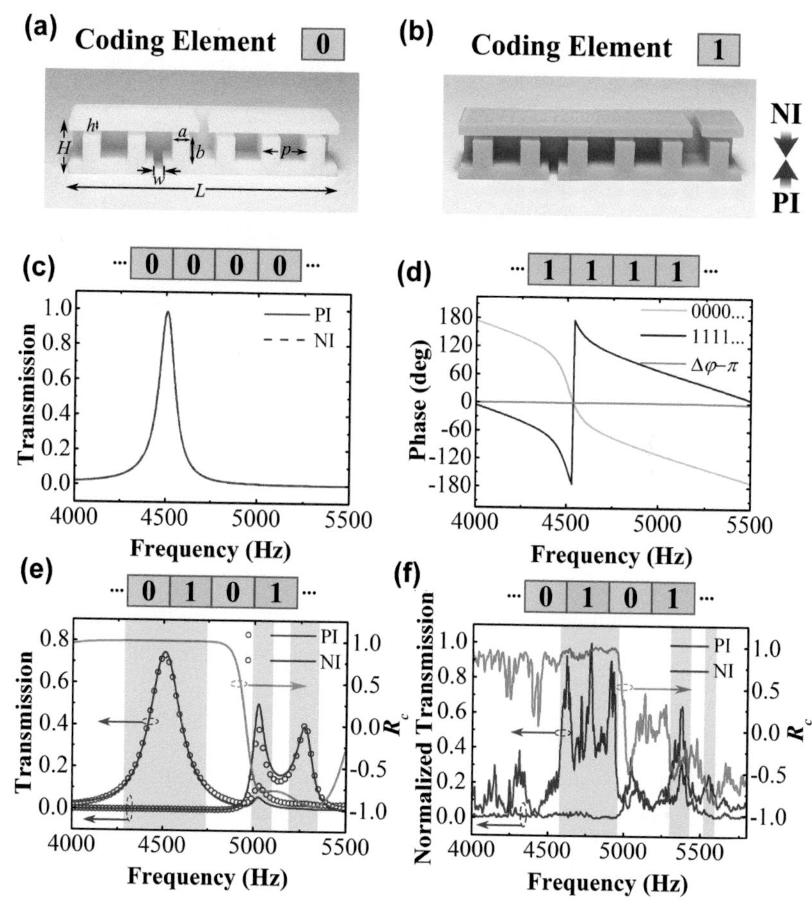

Figure 2.9 Multiband asymmetric transmission device based on acoustic coding metamaterial. (a,b) Structure of the coding elements "0" and "1", respectively. (c,d) The amplitude and phase of the coding sequences "0000 . . . " and "1111 . . . ", respectively. (e,f) The simulated and experimental results of the transmission spectra for both PI and NI, respectively.

absorbent material. A Brüel & Kjær (B&K) Type 4206 impedance tube was employed to mimic point source excitation and placed 1 m away from the sample. A microphone (B&K Type 4187) was used to measure the sound pressure of the transmitted wave. The measured transmission spectra of the sample is provided in Figure 2.9f, where three asymmetric bands can be found at around 4,700, 5,380 and 5,560 Hz, with little blue shift relative to the simulation results.

The acoustic pressure distribution is shown in Figure 2.10 to further investigate the working mechanism of the acoustic coding metamaterial in achieving

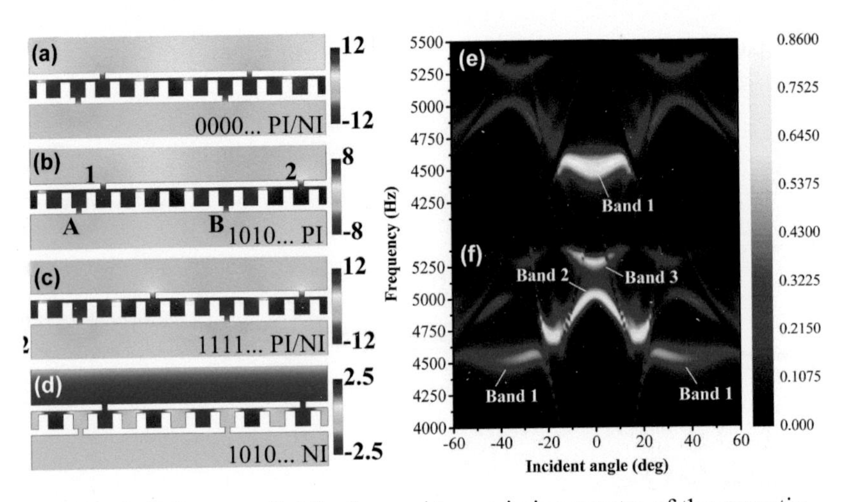

Figure 2.10 Pressure distribution and transmission spectra of the acoustic multiband asymmetric transmission device. (a,c) Pressure distributions of the "0000 . . . " and "1111 . . . " sequences for both PI and NI, respectively. (b,d) Pressure distributions of the "1010 . . . " sequence for the PI and NI, respectively. (e,f) The calculated transmission spectra as a function of incident angle.

asymmetric transmission. Almost identical acoustic pressure distribution can be observed in the pure "0000 . . . " (see Figure 2.10a) and "1111 . . . " (see Figure 2.10c) sequences for both PI and NI, which indicates a symmetric transmission. As the pressure at the input and output slits is in phase for both sequences, we would expect a destructive interference of acoustic wave for the NI (see Figure 2.10d) and a constructive interference for the PI (see Figure 2.10b), which leads to the phenomenon of asymmetric transmission.

The authors also evaluated the performance of the phenomenon of multiband asymmetric transmission under oblique incidence. Figures 2.10e and 2.10f provide the calculated transmission spectra for incident angle from 0° to 60°. For the PI case, only one obvious transmission band can be observed at 4,508 Hz, which is due to the lower-order resonance and is sensitive to the incident angle. However, multiple transmission bands can be observed for the NI case. Band 1 for NI at around 4,508 Hz is the reciprocal band of band 1 for PI, as can be verified from the incident angle that is identical to the angle range for PI. For larger incident angles, the transmission for NI gradually increases to 0.3, while it reduces to almost zero for PI, resulting in a shift of contrast radio from 1 to −1. Bands 2 and 3 can be observed in Figure 2.10f at larger incident angles, which is consistent with numerical simulations and experimental measurements. Taking advantage of the broadband operation, high-contrast ratio and subwavelength

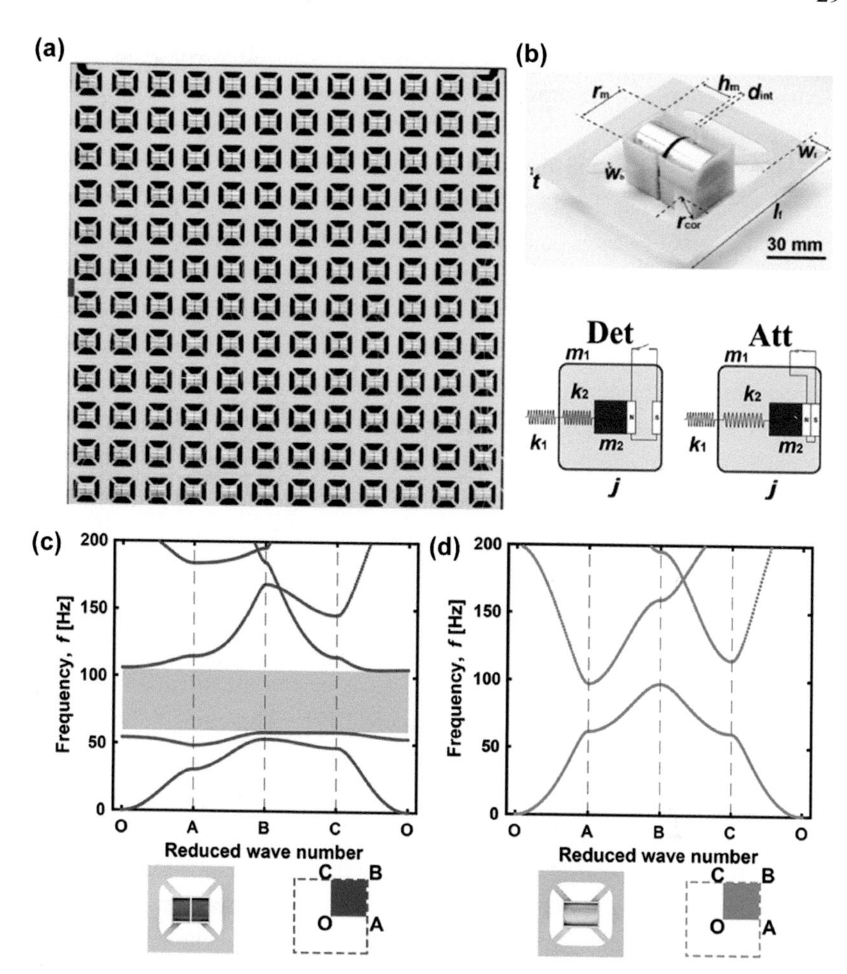

Figure 2.11 Digital elastic metamaterial with programmable functionalities. (a) Fabricated sample of the proposed digital elastic metamaterial. (b) Structure of the unit cell. The lower two schematics show the internal view of the unit cell under the Det and Att modes. (c,d) The dispersion curves of unit cell at the Att and Det modes, respectively.

thickness, the authors note this acoustic coding metamaterial featuring asymmetric transmission can be used as acoustic rectifiers, and it may promise ultrathin unidirectional devices in the future.

Despite the emergent acoustic coding metamaterials of recent years, it is challenging to manipulate low-frequency acoustic waves in broadband using traditional elastic metamaterials, which feature unusual properties such as negative effective mass density [167], negative effective modulus [168], negative Poisson's ratio [169], etc. Wang et al. proposed a digital elastic metamaterial with programmable functionalities such as wave guiding and wave isolation, by electrically controlling the current of electromagnets embedded in each resonator of the unit cell.

Figure 2.11a shows the fabricated sample of the proposed digital elastic metamaterial that has 12×12 unit cells. Each unit cell consists of a square frame, two pairs of beams, and two electromagnets connected in a series. The internal resonator with mass m_2 = 88 g is connected with the outer structure through a spring k_2 = 135 N m^{-1}. The electromagnets in each unit cell can attach or detach from each other, corresponding to the attaching (Att, see Figure 2.11b) and detaching (Det, see Figure 2.11b) modes, when the current is switched to ON and OFF, respectively. For example, when the current is off, the resonator can oscillates freely in the unit cell at the resonant frequency f_R = 6.23 Hz. While the current is switched on, the two electromagnets attach to each other due to the strong EM force.

Figures 2.11c and 2.11d provide the dispersion curve of the unit cell at the Att and Det modes, respectively. An obvious bandgap is observed from 59.5 to 106.1 Hz for the Det mode (see Figure 2.12a). The lower limit of the bandgap (i.e. the second branch) is due to the local resonance of the internal resonator, which disappears for the Att mode (see Figure 2.12b). Therefore, the structural transformation enables the switch of bandgap mode – that is the transmission state in the Att mode and the prohibit state in the Det mode. The two distinct states can therefore be designated with binary bits 0 and 1, enabling the first experimental realization of the acoustic programmable metamaterial.

The spatial mixture of the acoustic coding particles produces an extra flexibility in engineering the bandgap of the acoustic programmable metamaterial. Figures 2.12a–j show the frequency ranges of the bandgaps of all possible sequences for a 3×3 super cell as the number of 0 bits increases from 0 to 9. Note that only 26 sequences are independent from all the 2^9 sequences due to the structural symmetry. The bandgap can be freely designed by changing the number of element "0" in the super cell. For example, when the super cell is filled with all "1" bits, there is no bandgap. The frequency range of the bandgap increases from 0 to 45 Hz monotonically as the number of 0 bits increases from 0 to 9. The distribution of "0" and "1" bits in the super cell also affects the frequency range of the bandgap.

Experiments were conducted to demonstrate the waveguiding behavior of the acoustic programmable metamaterial. The experimental configuration is described briefly as follows. The whole structure was suspended with strings linking the upper edge, and the plane was aligned with the gravitational direction. An electrodynamic shaker (HEV-50, Nanjing Foneng, China, indicated by the red line in Figure 2.11a) was applied to the lattice edge to produce the

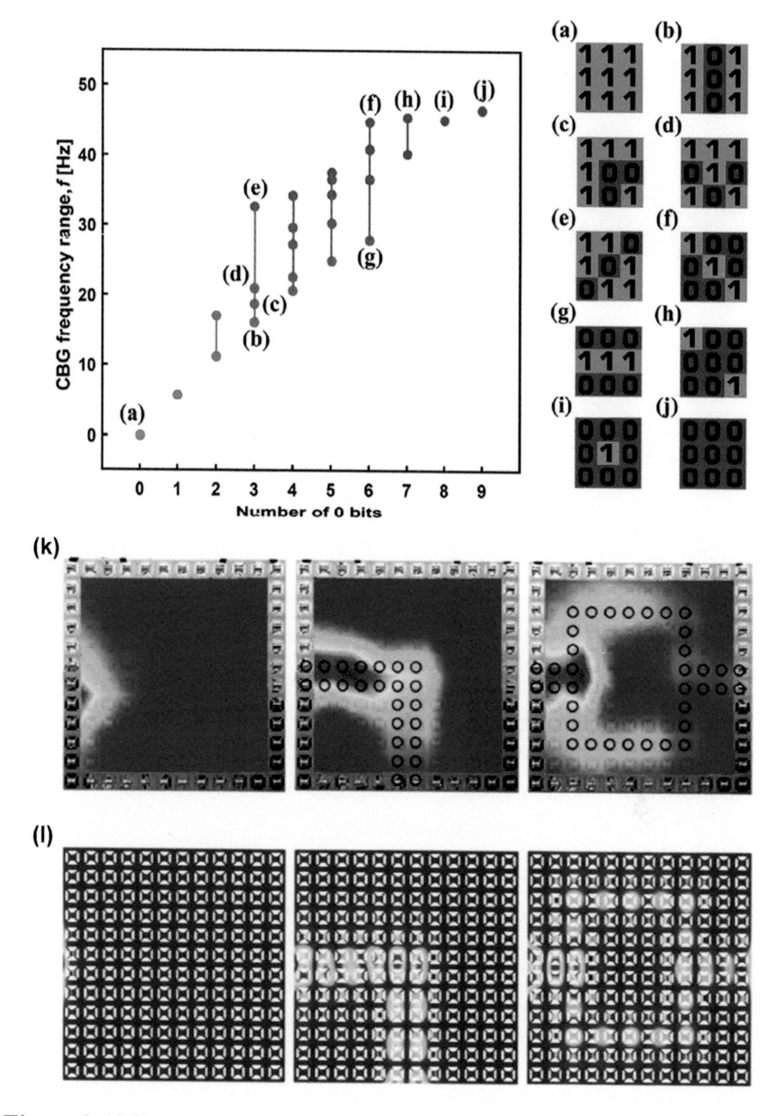

Figure 2.12 Frequency ranges of bandgaps and waveguiding behavior of the acoustic programmable metamaterial. (a–j) Frequency ranges of bandgaps of all possible sequences for a 3×3 super cell as the number of 0 bits increases from 0 to 9. (k,l) The measured and simulated distributions of the velocity field at 60 Hz for three different coding patterns.

continuous out-of-plane harmonic excitation. The wave-propagating field was recorded by a scanning laser Doppler vibrometer (PSV-400, Polytec, Inc., Germany). Three different coding patterns were measured in the experiment: all cells in Det mode, an L-shaped route, and a bypass route.

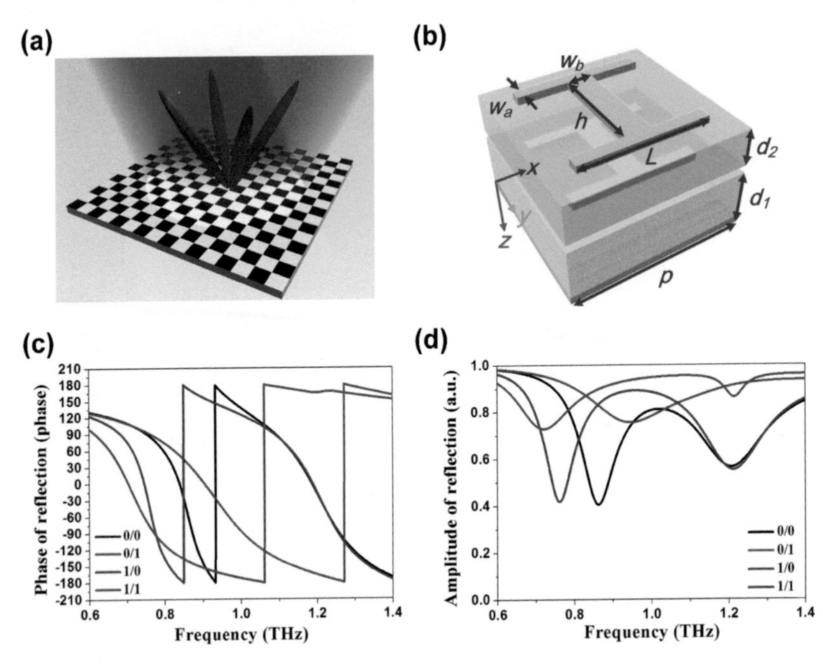

Figure 2.13 Dual-band bifunctional coding metamaterial. (a) Schematic illustration of the dual-functional behavior of the dual-band bifunctional coding metamaterial. (b) Structure of the coding particle. (c,d) Reflection phases and amplitudes of the four coding particle "0/0", "0/1", "1/0" and "1/1", respectively.

Figures 2.12k and 2.12i show the measured and simulated distributions of the velocity field at 60 Hz for the three different coding patterns. For the all Det mode coding patterns, the field was blocked at the input point, while for the other two coding patterns, the acoustic wave could propagate along the route filled by the "1" elements. Excellent agreement can be observed from the simulation and experimental results.

This technique is different from previously reported tunable methods such as piezo shunting [170–172], structural deformation [173,174], and shape memory effect [175,176]. However, the frequency tuning approaches of these designs are rather sophisticated, making it difficult to realize a digital acoustic element that is dynamically tunable, particularly at low frequency.

2.5 Dual-Band Bifunctional Coding Metamaterial

All the aforementioned coding metamaterials possess a single functionality for a given coding pattern because the digital state of each constituent coding particle is fixed at the designed frequency. As the reflection (or refraction) coefficient of a coding particle varies as a function of frequency due to the dispersion nature of the real structure, it is possible to encode two independent digital states into a single coding particle at two different frequencies, enabling bifunctional and multifunctional coding metamaterials.

To illustrate this concept, we designed a frequency-dependent dual-band coding metamaterial at THz frequency [177]. Figure 2.13a schematically illustrates the dual-functional behavior of the dual-band coding metamaterial, in which the low-frequency (see the red beam) light is deflected to a certain plane, while the higher-frequency light (see the blue beam) is redirected to the orthogonal plane. The dual-functional performance comes from the orthogonality of different frequency bands, which allows us to encode different coding patterns – i.e. different functionalities – to a single coding metamaterial.

Figure 2.13b is the real structure for implementing the dual-band coding metamaterial at THz frequency. In order to have independent reflection responses at two different frequencies, two electric-LC (ELC) structures with independent geometrical parameters were designed on top of two polyimide layers. Similar to all the reflection-type coding metamaterial, a metallic ground sheet is required at the back side of the bottom polyimide layers to guarantee zero transmission and close-to-unity reflection. Four different coding particles – "0/0", "0/1", "1/0" and "1/1" – were optimized through numerical simulations to constitute the 1-bit dual-band coding metasurface. The code before and after the slash indicates the digital states at the lower (0.78 THz) and higher (1.19 THz) frequencies, respectively.

Figures 2.13c and 2.13d show the simulated reflection phases and amplitudes of the four coding particles from 0.6 to 1.4 THz. It can be observed that the phase difference is about 177° at the lower frequency and 165° at the higher frequency, which satisfies the opposite phase response of the 1-bit coding metasurface. The amplitude of reflection inevitably suffers from absorption at resonant frequencies, which is mainly attributed to the dielectric loss of the polyimide layer. We remark that the reflection amplitude could be significantly improved by employing some low-loss materials as the substrate, such as crystal quartz, which has a tiny loss tangent of 0.0004 at 1 THz.

Three different examples are given to demonstrate the bifunctional performance of the dual-band coding metamaterial. The first coding pattern is composed of two sub-coding sequences – "00001111 . . . " and "000111 . . . " – at the lower and higher frequencies, respectively, with mutually orthogonal varying direction. Figures 2.14a and 2.14d show the simulated radiation patterns of the first coding pattern at the lower and higher frequencies,

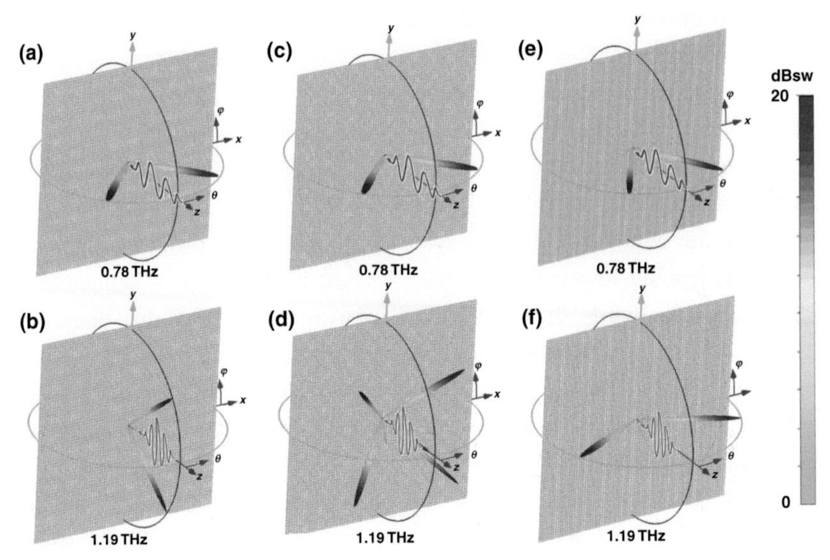

Figure 2.14 Simulated radiation patterns of the dual-band coding metamaterial at the (b,d,f) lower frequency and (a,c,e) higher frequency. (a,b) "00001111 . . . " varying along the x direction at the lower frequency, "000111 . . . " varying along the y direction at the higher frequency. (c,d) "00001111 . . . " varying along the x direction at the lower frequency, chessboard coding pattern at the higher frequency. (e,f) "0000011111 . . . " varying along the x direction at the lower frequency, "0011 . . . " varying along the y direction at the higher frequency.

respectively. It is clearly observed that the normal incident beam is split into two beams in the x-z plane at the lower frequency and into two beams in the y-z plane at the higher frequency.

For the second coding pattern, we changed the sub-coding sequence at the higher frequency to the chessboard distribution. It can be found from Figures 2.14c and 2.14d that the radiation pattern remains unchanged at the lower frequency and becomes a four-beam radiation pattern at the higher frequency. The highly consistent radiation pattern at the lower frequency for the first and second coding patterns (see Figures 2.14a and 2.14c) demonstrates excellent isolation between the codes at the two designed frequencies.

The third coding pattern shows an interesting phenomenon, as can be observed from Figures 2.14e and 2.14f, where the deflection angle at the lower frequency is smaller than that at the higher frequency. This is counterintuitive against Snell's law, as the deflection angle always decreases with the increase of frequency. Owing to the independent coding states at the lower and higher frequencies, this unusual phenomenon can be realized by setting the

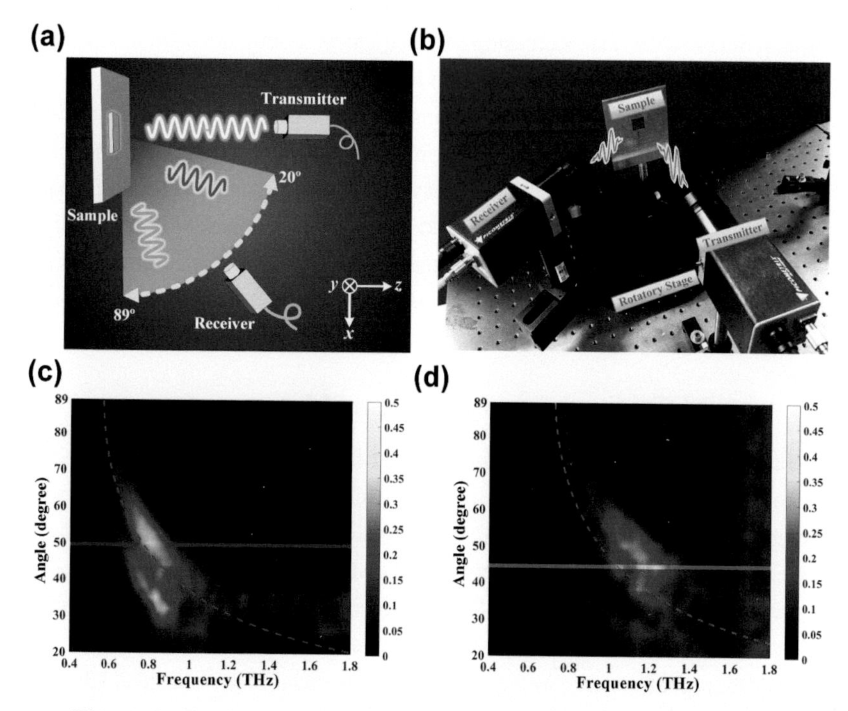

Figure 2.15 Experimental characterization of the dual-band coding metamaterial. (a,b) Schematic and photo of the experimental setup. (c,d) The reflection spectra measured at the *x-z* plane and *y-z* plane, respectively.

coding sequences at the lower and higher frequencies as "0000011111 . . . " and "0011 . . . ", respectively. By normalizing the radiation pattern to the reference case, which is a bare metallic sheet with the same size, the efficiency is estimated to be around 40–50%.

A rotary THz-TDS was employed to characterize the performance of the fabricated sample, as illustrated in Figures 2.15a and 2.15b. Two photoconductive antennas (TR4100-RX1, API Advanced Photonix, Inc.), excited by a commercial ultrafast erbium fiber laser system (T-Gauge, API Advanced Photonix, Inc.) that works from 0.3 to 3 THz, were used as the transmitter and receiver. The transmitter illuminates the sample normally with a 5 mm-wide THz beam, and the receiver measures the reflection spectra from 20° to 89° with a step of 3°. The direct transmission when the transmitter and receiver were coaxially aligned was recorded as the reference signal.

Figures 2.15c and 2.15d are the reflection spectra measured at the *x-z* plane and *y-z* plane, respectively. One can easily observe a reflection peak around 50° at 0.8 THz (see Figure 2.15c), and a reflection peak around 44° at 1.2 THz (see Figure 2.15d), which is close to the simulation result. Note that the reflection angle reduces with the increasing of frequency, being consistent with the theoretical results marked by the dotted line.

The future frequency-dependent coding metasurfaces should advance toward higher-bit and multiband, which have more freedom in controlling the THz wave, by elaborately optimizing multilayered structures with several resonant frequencies within the frequency band of interest. Such multiband coding metasurfaces can help enhance the transmission rate of wireless communication systems, and can be exploited to realize color holography if this concept is extended to visible light.

2.6 Transmission-Type Coding Metamaterials

To adapt to various applications with different feeding sources, one may need to design reflection-type coding metasurfaces, while in other cases it may be more desirable to manipulate the transmitted wave. Compared with the transmission-type coding metasurfaces, the reflection-type coding metasurfaces are much easier to design and fabricate for the following reasons. First, one can easily obtain a close-to-unity reflection amplitude for the reflection-type coding metasurface at any reflection phase, owing to the existence of the metallic ground sheet added at the back side of the dielectric substrate. Second, the fabrication process for the reflection-type coding metasurface is typically simpler than that of the transmission-type, especially at the THz frequency, where the sample is usually required to be peeled off from the carrier substrate. Things are different for the transmission-type coding metasurface, mostly because of the difficulty in designing an efficient coding particle with a high transmission coefficient at any transmission phase.

In 2013, Alù's group pointed out the general limitations on the attainable phase coverage and amplitude for a single, ultrathin metasurface [98]. The group stated that, for a single-layered, nonmagnetic metasurface, the transmitted amplitude and phase of the co-polarized component are not independent. The transmission phase is fundamentally limited in the range of $-\pi/2 \sim \pi/2$, with the amplitude approaching zero close to the bound of this range. The manipulation of the cross-polarized component is also not optimistic. The transmitted amplitude and phase are dependent on each other, and the maximum attainable cross-coupling is limited to 0.5. For these reasons, it is impossible to achieve an arbitrary phase without sacrificing the amplitude, for controlling either the co-polarized or the cross-polarized transmitted wave.

To overcome this limitation, Alù and colleagues proposed to use a three-layered structure, with the inner and outer layers having independent surface effective reactances. In this configuration, the transmitted phase could shift from 0° to 360° with almost unity amplitude. Corresponding surface reactances for the three layers were calculated to realize the effect of beam bending and beam focusing with almost 100% efficiency.

In the same year, Professor Grbic's group proposed to realize a reflection-less sheet using only two metallic layers, one for exciting the electric responses and another for magnetic response. As the mechanism was inspired by the concept of Huygens's surface, they called this metamaterial Huygens's surface. By assuming the desired wave distribution on both sides of the metasurface, the electric sheet admittance and magnetic sheet impedance can be analytically calculated, which serves as a guide for the structural design of the two metallic layers.

Although this approach could also bend and focus the EM wave with almost 100% efficiency, the plane of the metallic structure is parallel to the propagation direction, which is very challenging to fabricate at THz frequency using state-of -the-art micro-fabrication techniques. One of the advantages of the transmission-type metasurface is the convenience of the feeding source, which can be placed at the transmitted side of the metasurface to avoid the blockage problem encountered in the reflection-type metasurface. The total volume of the transmission-type metasurface is also smaller compared with the reflection-type metasurface, leading to a more compact design.

Before we started to design the transmission-type coding metasurface, we made a brief review of a gradient-phase transmit array designed to manipulate the THz waves using single-layered metasurfaces [178,179]. They can take full control of the phase and amplitude of transmission by varying the opening angle and orientation of each C-shaped resonator. Unfortunately, restricted by the single-layered configuration, the maximum amplitude of transmission was limited to 0.4, which is well below the levels required in most practical applications. This value was even lower for the real sample due to the impedance mismatch and Fabry-Perot resonance that resulted from the thick silicon substrate with high permittivity. Therefore, it was necessary for us to control the THz waves efficiently using a transmission-type coding metasurface [146]. Here, we developed a free-standing transmission-type coding metasurface at the THz frequency, as illustrated in Figure 2.16a, which could convert the incident wave to its cross-polarization and bend it to an anomalous direction.

Figures 2.16b and 2.16c display 3D and 2D views of the structure for the transmission-type coding metasurface, which consists of three SRRs separated by two polyimide spacers with 40 um thickness. The opening angles of the bottom, middle and top SRRs were 0°, 45° and 90° with respect to the y axis. The rotational twist of the three SRRs produced magnetoelectric coupling that made a 90° polarization rotation of the illuminating wave. By tuning the opening angle of SRRs, we obtained eight coding particles used to build the 3-bit transmission-type coding metasurface.

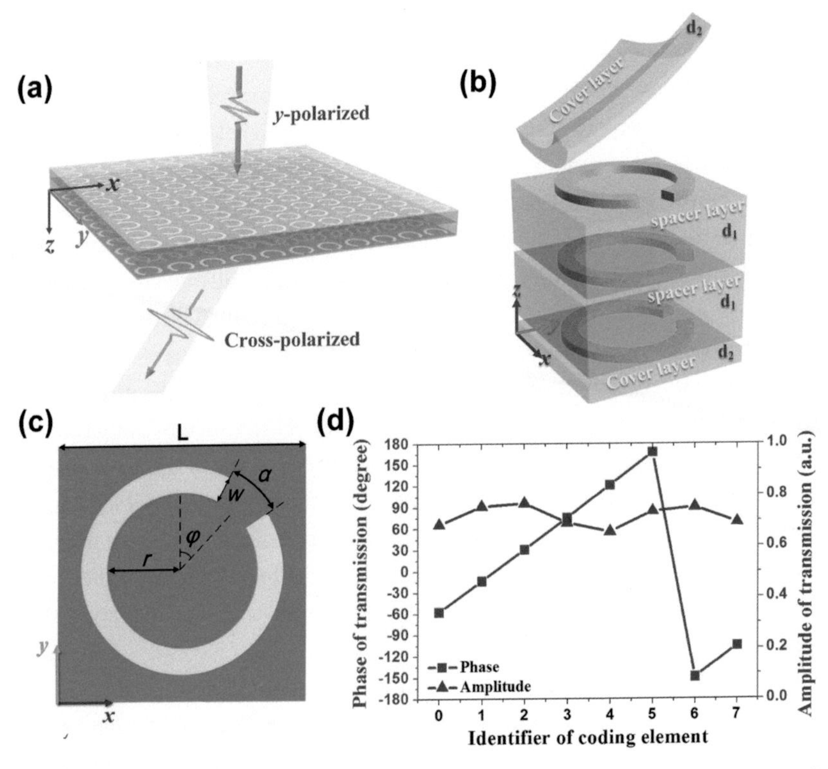

Figure 2.16 Transmission-type coding metamaterials at terahertz frequency. (a) Schematic illustration of the anomalous refraction of the THz transmission-type coding metamaterial. (b,c) perspective view and top view of the structure of the unit cell, respectively. (d) Transmission phase and amplitude of the eight coding particles at the designed frequency of 0.97 THz.

Figure 2.16d shows the amplitudes and phases of the eight coding particles at the designed frequency of 0.97 THz. Each adjacent coding particle has a phase step of 45°, and the amplitudes of all coding particles range from 0.65 to 0.76, exceeding the theoretical upper bound supported by the single-layered metasurface. Note that the structure of the last four coding particles can be readily obtained by rotating the orientation angle of the SRRs of the first four coding particles by 90°.

The performance of the transmission-type coding metasurface was evaluated by simulating the near-field distributions of the beam bending and Bessel beam focusing effects. Figure 2.17a shows the reflected wave when the plane wave is incident on the coding metasurface with a gradient coding sequence "0 1 2 3 4 5 6 7 . . . ". It can be observed that the refracted wave is propagating along a 21° angle with respect to the normal axis, perfectly agreeing with the theoretical value of

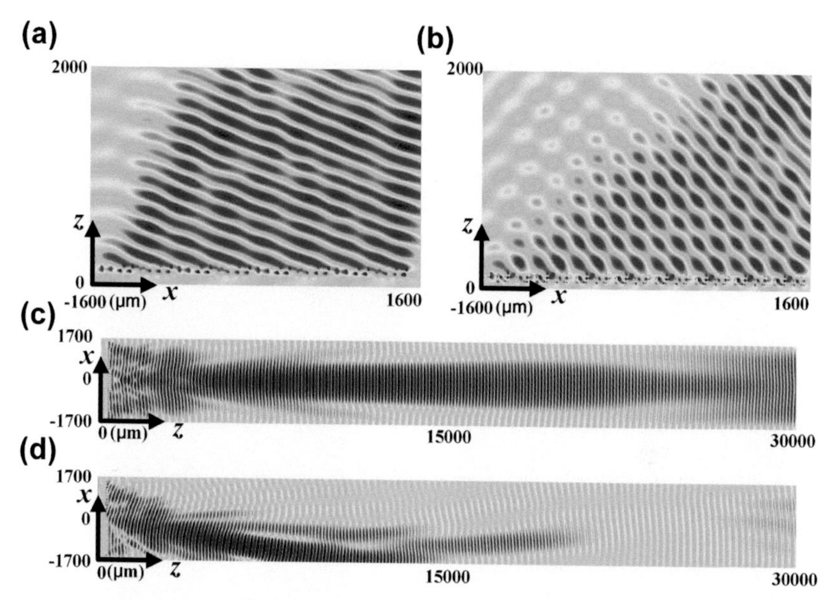

Figure 2.17 Simulation results of the THz z transmission-type coding metamaterial in refracting and focusing the THz wave. (a,b) Electric field distributions of the reflected wave for the gradient coding sequences "0 1 2 3 4 5 6 7 . . . " and "1 3 5 7 1 3 5 7 . . . ", respectively. (c,d) Electric field distributions for the Bessel beam generation at the normal and oblique directions, respectively.

21.13°. As the gradient coding sequence reduces to "1 3 5 7 1 3 5 7 . . . ", the refraction angle increases to 46°, as is shown in Figure 2.17b. The perturbances observed in both near-field distributions and attributed to the side lobes resulted from the undesired EM couplings and discontinuities of the discrete phase profile.

As the transmission-type coding metasurface has no shielding effect, we can demonstrate the function of Bessel beam forming with a gradient coding sequence that varies along the radial direction. A non-diffractive beam can be clearly identified from the simulated near-field distribution in Figure 2.17c, which was generated by normally illuminating the coding metasurface with a plane wave. The direction of the Bessel beam can be arbitrarily controlled by adding another gradient coding sequence that varies along the x direction, as demonstrated in Figure 2.17d. One can also manipulate the focusing distance by changing the period of the gradient coding sequence.

One of the difficulties in realizing transmission-type metasurfaces at the THz frequency is the fabrication of a free-standing sample without a rigid substrate, as shown in Figures 2.18a and 2.18b. The fabrication process is similar to that described in Section 2.1.2 for the reflection-type sample. After completing all

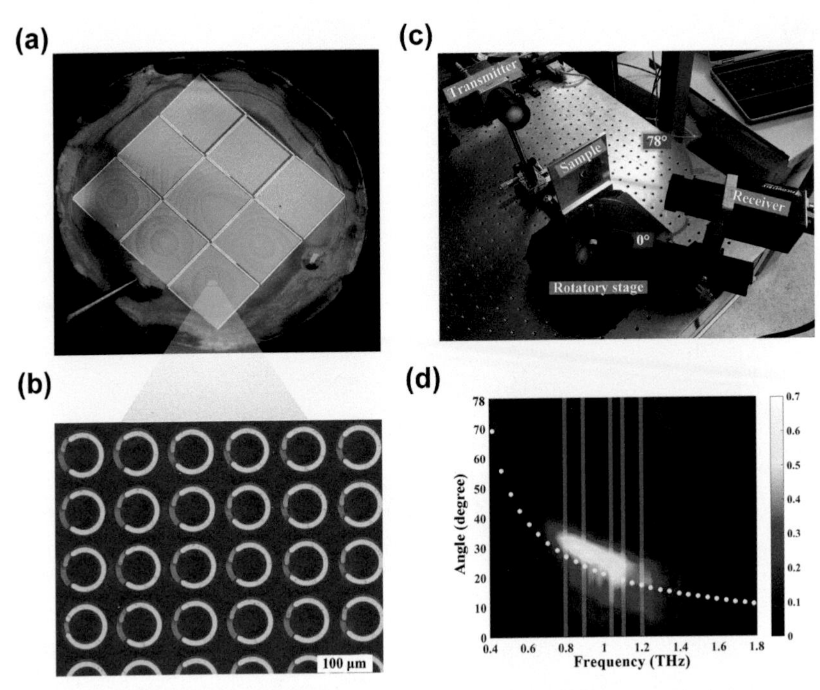

Figure 2.18 Experimental characterization of the THz transmission-type coding metamaterial. (a,b) Photo and microscopy image of the fabricated sample. (c) Photo of the experimental setup. (d) The transmission spectra of the fabricated sample with the gradient coding sequence "0 1 2 3 4 5 6 7 . . . ".

metallic and polyimide layers, we peeled off the sample from the silicon wafer by rinsing it in pure hydrofluoric acid (HF concentration > 40%) solvent for 30–60 minutes. The sample was robust against mechanical scratches and chemical corrosion, owing to the 5 um-thick polyimide layer covered at both sides of the sample. The flexibility of the free-standing sample should promise many interesting applications that require the samples to be perfectly attached to the object's surface.

The same rotary THz-TDS (see Figure 2.18c) was employed to characterize the transmission spectra of the fabricated sample with the gradient coding sequence "0 1 2 3 4 5 6 7 . . . ". The sample was attached to a sample holder with a square window in the middle. The receiver rotated from 0° to 78° with 3° steps to measure the transmission spectra from 0.4 to 1.8 THz, as shown in Figure 2.18d, where an obvious peak from receiving angles of 20° to 32° can be clearly observed in the frequency range from 0.8 to 1.2 THz. At the designed frequency of 1.04 THz, the transmission intensity reached the maximum at the receiving angle of 26°, which is slightly larger than the simulation result.

3 Anisotropic Digital Coding Metamaterials

3.1 Uniaxial Anisotropic Digital Coding Metamaterials

For the square patch design adopted in the initial coding metamaterial, the digital state was single-valued and therefore not relevant to the polarization direction. Hence, it falls into the category of isotropic coding metamaterial. Similar to the dual-band coding metamaterial, which utilizes the orthogonality of different frequencies, we can also design a polarization-dependent coding metamaterial that exploits the orthogonality between the x and y polarizations. Such polarization-dependent coding metamaterials can exhibit distinct functionalities under the x and y polarized incidences, and they are therefore classified as anisotropic coding metamaterials.

Figure 3.1a schematically illustrates the working mechanism of anisotropic coding metamaterials. The central anisotropic coding pattern can be

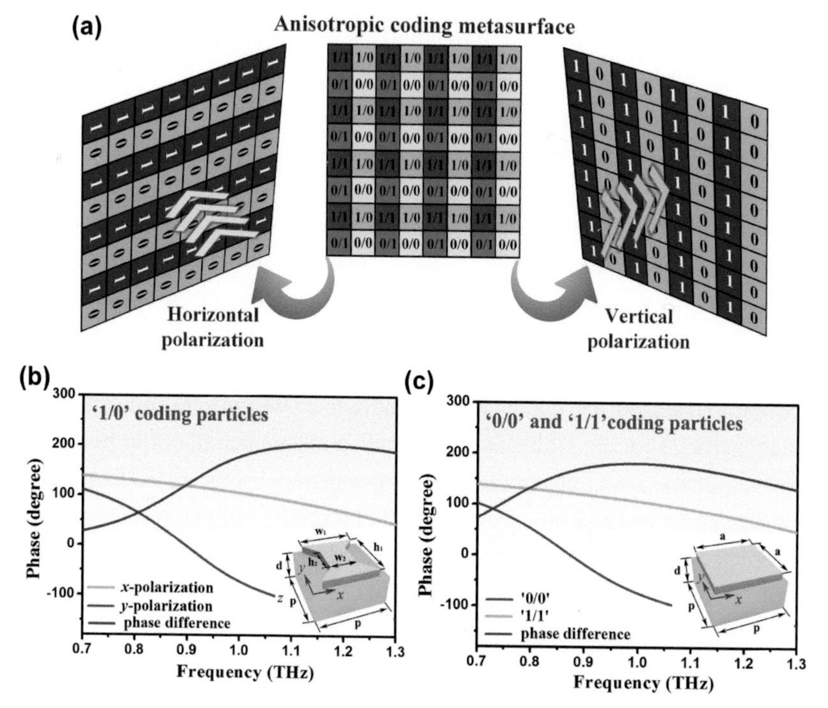

Figure 3.1 Uniaxial anisotropic coding metamaterial. (a) Schematic illustration of the polarization-dependent behavior of the uniaxial anisotropic coding metamaterial. (b) Reflection phases of the "1/0" anisotropic coding particle under the x and y polarizations. (c) Reflection phases of the "0/0" and "1/1" isotropic coding particles.

expressed by a 2×2 matrix [0/0, 0/1; 1/0, 1/1], in which the codes before and after the slash represent the digital states under the horizontally and vertically polarized illuminations, respectively. For the horizontally polarized incidence (see the left panel of Figure 3.1a), the coding pattern is "0 1 0 1 . . . " varying along the vertical direction, which deflects the incident beam to two beams in the vertical plane, while for the vertically polarized incidence, the coding pattern becomes "0 1 0 1 . . . " varying along the horizontal direction, which results in two-beam radiation in the horizontal plane. The functionality of this anisotropic matrix is numerically simulated in Figures 3.2a and 3.2b.

Each code in the matrix is composed of a super unit cell with a size of 4×4. It can be seen that the normally incident beams with x and y polarizations are split into two beams located in the y-z plane and the x-z plane, respectively. When the incident beam is polarized 45° with respect to the x or y axis, four beams will be generated, with two beams containing the x polarization and another two beams containing the y polarization, making it a spatial beam splitter.

To be accurate, the reflection (refraction) coefficient of an anisotropic coding metamaterial can be expressed by a tensor $\overline{R}_{mn}$,

$$\overline{R}_{mn} = \begin{bmatrix} \hat{x}R^x_{mn} & 0 \\ 0 & \hat{y}R^y_{mn} \end{bmatrix} \tag{3.1}$$

where R^x_{mn} and R^y_{mn} are the reflection (refraction) coefficients of a certain coding particle (mn) on the coding metamaterial under the x and y polarizations, respectively. Apparently, $R^x_{mn} = R^y_{mn}$ for the isotropic coding metamaterial.

To implement the anisotropic coding metamaterial with real structure, both isotropic and anisotropic coding particles have to be designed to construct the 4 coding particles for the 1-bit anisotropic coding metamaterial, or the 16 coding particles for the 2-bit anisotropic coding metamaterial. Figures 3.1b and 3.1c show the structures and broadband reflection responses of the anisotropic and isotropic coding particles, respectively. A dumbbell-shaped structure, characterized by heights h_1 and h_2 and widths w_1 and w_2, can have independent reflection phases under the x- and y-polarized incidences.

To further demonstrate the versatile behavior of the anisotropic coding metamaterial, another example is given in Figures 3.2c and 3.2d. The code under the x polarization stays the same as that in Figure 3.2a, while the code under the y polarization changes to a random coding pattern. In this scenario, the radiation pattern under the x polarization remains two beams, but it becomes the random diffusion pattern under the y polarization. The high consistency between the radiation pattern in Figures 3.2a and 3.2c demonstrates the

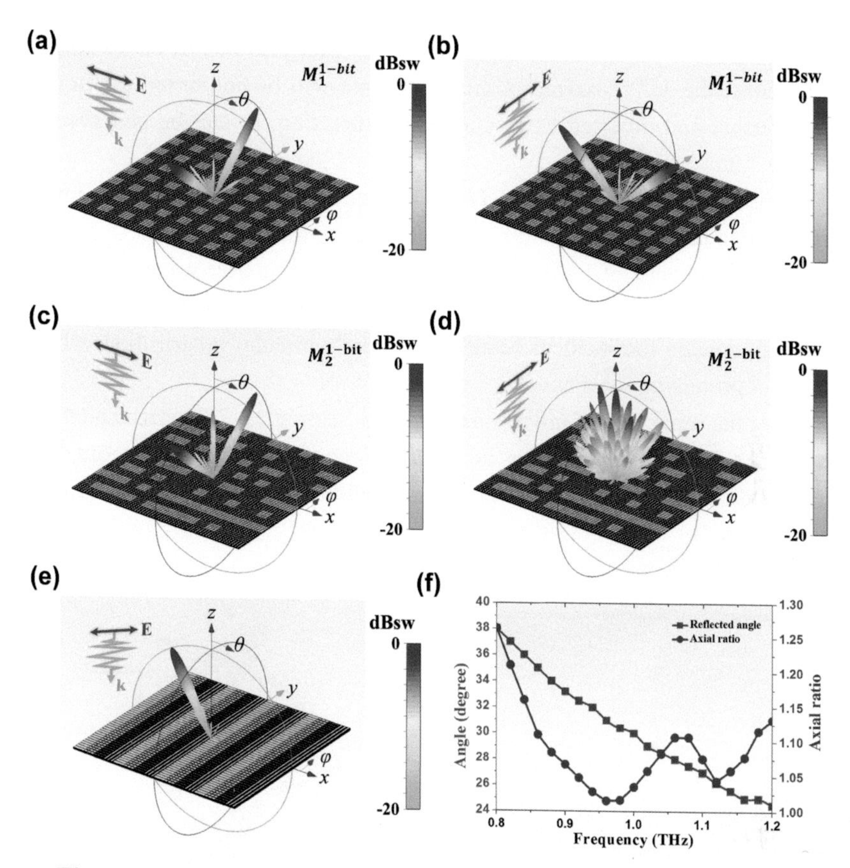

Figure 3.2 Simulated radiation patterns for different anisotropic coding patterns. (a,b) 3D radiation patterns of the anisotropic coding matrix [0/0, 0/1; 1/0, 1/1] under the *x* and *y* polarizations, respectively. (c,d) 3D radiation patterns of the anisotropic coding matrix M_2 (the coding sequence under *x* polarization is "0101 . . . " varying along the *y* direction, the coding sequence *y* polarization is random sequence) under the *x* and *y* polarizations, respectively. (e) 3D radiation patterns of the reflection-type quarter wave plate. (f) The axial ratio and reflection angle of the reflection-type quarter wave plate from 0.8 to 1.2 THz.

excellent isolation between the functionality of the anisotropic coding metamaterial under orthogonal polarizations.

We could realize more functionalities with a 2-bit anisotropic coding metamaterial. For example, the normal incidence can be redirected to a certain direction under the *x* polarization, but to another direction under the *y* polarization. More interestingly, a reflection-type quarter-wave plate can be designed with the coding sequence "00/01 01/10, 10/11, 11/11 . . . ". As the sub-coding sequences under both polarizations grade along the same direction, and as each code under the x and y polarizations has a 90° phase difference, the 45° polarized normal incidence will be converted to circular polarization and diverted to an anomalous direction, as can be observed in Figure 3.2c.

The axial ratio from 0.8 to 1.2 THz is plotted in Figure 3.2d, which is lower than 1.26 in the entire bandwidth and reaches the minimum value of 1.03 at the operation frequency of 1 THz. As this reflection-type quarter-wave plate could generate a circularly polarized beam in an oblique direction, the reflected wave does not contain the incident beam and is purely circular polarization, which makes it promising for many THz applications.

Using the same lift-off process described in Section 2.1.2, we fabricated the sample in Figure 3.3a. In this fabrication process, because only gold was deposited on the silicon wafer, the sample could be easily peeled off from the

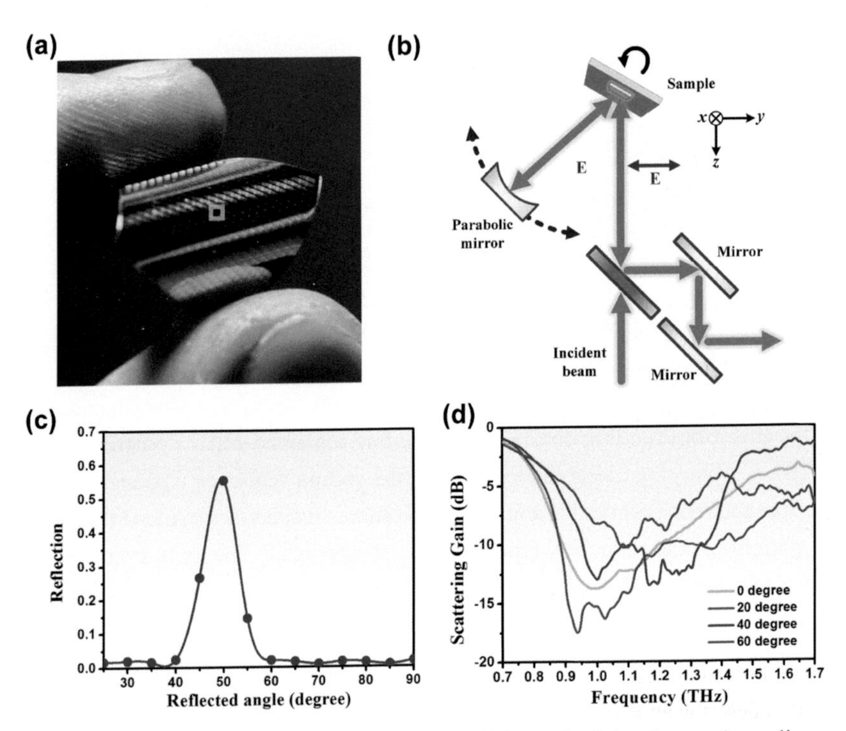

Figure 3.3 Experimental characterization of the uniaxial anisotropic coding metamaterial at terahertz frequency. (a) Photo of the fabricated sample. (b) Schematic of the experimental setup. (c) The measured anomalous reflection spectra of the sample from 25° to 90° under the x-polarized incidence. (d) The measured RCS reduction of the sample under the y-polarization at four different incident angles 0°, 20°, 40°, and 60°.

silicon wafer due to the poor adhesion between gold and silicon. As shown in Figure 3.3a, the sample features excellent flexibility and can be easily bent and twisted. Figure 3.3c shows the anomalous reflection spectra measured by the same rotary THz-TDS introduced in Section 2.5. A peak appears at the center angle of 51° with normalized amplitude of around 0.56.

Figure 3.3d shows the curves of RCS reduction at different incident angles, which were measured by a theta-to-theta THz-TDS (Zomega Z3, East Greenbush, NY, USA), as schematically illustrated in Figure 3.3b. The sample was attached to a holder and located on a rotary stage that could rotate from 20° to 90° automatically. A parabolic mirror, automatically aiming at the specular direction of the incidence, reflected the THz beam back to the sample. The signal was finally collected by the receiving photoconductive antenna through several reflecting mirrors. As the received THz signal was reflected by the sample twice, the square root of measured data should be adopted as the real refection coefficient.

Figure 3.3d shows the RCS reduction of the sample under the orthogonal polarization at different incident angles of 0°, 20°, 40° and 60°, which were normalized to the specular reflections from the reference sample made of 200 nm-thick gold layer. Excellent RCS reduction was obtained in a relatively broad bandwidth. At the normal incidence, a −10 dB reduction was achieved from 0.9 to 1.2 THz, while a −10 dB reduction range was further expanded to 0.5 THz (from 0.88 to 1.38 THz) under 20° oblique incidence. For larger incident angles, the random diffusion effect deteriorates because the wave vector is no longer parallel to the structure plane.

Here, we would like to envision potential applications of the anisotropic coding metamaterial. At microwave frequency, it could help enhance the transmission rate of wireless communication systems by modulating independent signals on two orthogonal polarizations. It could also be extended to the visible light spectrum to increase the storage density of current storage media (such as light disk) by multiplexing 2-bit data into one storage dot. We could also encode two different images into one hologram to develop 3D holographic display techniques.

3.2 Full-Tensor Digital Coding Metamaterial

Phase, frequency, wave vector and polarization are necessary parameters commonly used to describe an EM device. For the anisotropic coding metamaterial described in Section 3.1, we aimed at manipulating only the phase profile of the co-polarized component of the illuminating wave. Hence, the off-diagonal elements in the tensor matrix should be close to zero. As polarization is another property of EM waves, it would further enhance the ability of the coding metamaterials if they could control the polarization [180]. Here, a rectangular-shaped SRR was designed to control both the phase and polarization of the illuminating wave. To be accurate, the following tensor matrix was used to describe the reflection/refraction properties of such a coding particle:

$$\overline{R} = \begin{bmatrix} R_{xx} & R_{xy} \\ R_{yx} & R_{yy} \end{bmatrix} \tag{3.2}$$

For both isotropic and anisotropic digital coding particles, the cross-polarized components R_{xy} and R_{yx} should be zero, while for the coding particle that could convert the polarization state of the EM wave, R_{xy} and R_{yx} should approach unity, and R_{xx} and R_{yy} should be minimized. As the EM property of such coding metamaterial was described by a tensor form, it was named a tensor coding metamaterial. Two different working modes were supported by the tensor coding metamaterial, which were the spatially propagating wave (PW) mode and the surface wave (SW) mode.

Figure 3.3a schematically illustrates the PW mode, in which the y-polarized normal incidence (the red beam) is anomalously reflected with cross-polarization (the green beam). For the SW mode, we remark that the tensor coding metamaterials support both TM- and TE-mode SWs, as illustrated in Figures 3.4b and 3.4c, respectively. As depicted in Figure 3.4b, as the y-polarized PW is normally incident on the tensor coding metamaterial, it was converted to the TM-mode SW propagating on the quartz substrate placed next to

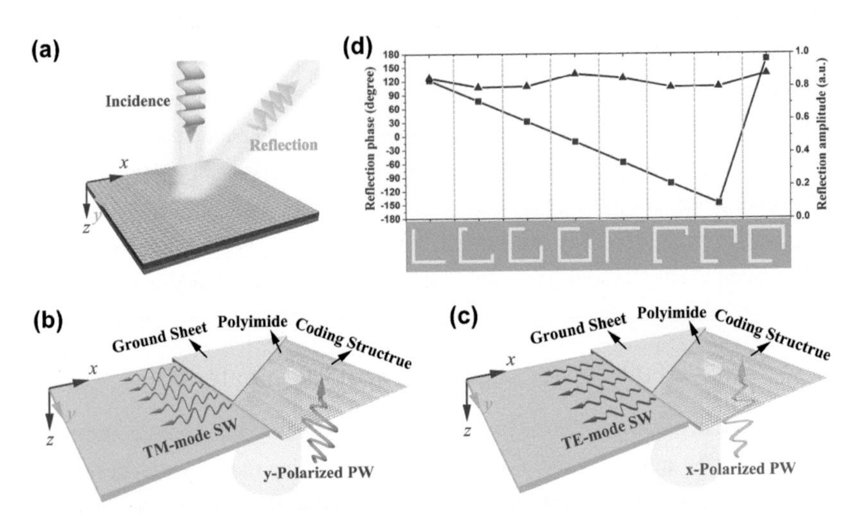

Figure 3.4 Tensor coding metamaterial. (a) Schematic illustration of the tensor coding metamaterial in controlling PW. (b,c) Schematic illustration of the tensor coding metamaterial in controlling TM-mode and TE-mode SWs, respectively. (d) Reflection phases and amplitude of the eight coding particles at the designed frequency of 0.93 THz.

the coding metamaterial. As for the x-polarized PW, it was converted to the TE-mode SW, as illustrated in Figure 3.4c. Note that this design differs from the PW-SW convertor using the H-shaped gradient metamaterial in the following aspects.[181]

First, the previous design only allows the x-polarized incidence to be converted to the TM-mode SW, whereas our design supports both x- and y-polarized PW incidences. Second, as the H-shaped structure can only manipulate the co-polarized component, then only the x-polarized PW is converted to the TM-mode SW. But for the tensor coding metamaterial with the polarization conversion ability, the x-polarized PW is converted to the TE-mode SW, and the y-polarized PW is converted to the TM-mode SW. Third, this work presents the first design and experimental demonstration of PW-SW at the THz frequency.

To realize these functionalities with high efficiency, eight SRR structures were designed to construct the 3-bit tensor coding particle, as given in the bottom panel in Figure 3.4d. At the intended frequency of 0.93 THz, each adjacent coding particle has a phase difference of 45°. The amplitude of cross-polarization exceeds 0.8 for all the eight coding particles, ensuring the efficient control of PW and SW when they are encoded in an array. Due to the geometrical symmetry of the structure with respect to the 45° axis, the last four coding particles can be simply obtained by rotating the first four 90°.

Numerical simulations were carried out to evaluate the PW-SW conversion effect of the tensor coding metamaterial. The first example demonstrates the conversion from the y-polarized PW to the TM-mode SW, using a coding sequence "1 1 3 3 5 5 7 7 . . . ". A waveguide port placed 600 μm above the coding metamaterial provided the required quasi-plane wave. To receive the converted SW for the subsequent measurement, an 80 um-thick crystal quartz substrate (permittivity $\varepsilon_r = 4.4$; loss tangent $\delta = 0.0004$) was placed next to the coding metamaterial. In contrast to polyimide, however, crystal quartz can exhibit much lower absorption loss, emerging as a suitable material for implement coding and programmable metamaterials at THz, infrared and visible light frequencies. Note that the thickness of quartz substrate was carefully selected to take full consideration of the good impedance match between coding metamaterial and quartz substrate, as well as the fabrication feasibility.

The simulated electric field distribution (E_x component) at 0.75 THz is shown in Figure 3.5a, where TM-mode SW is clearly observed on the quartz substrate. One may notice that the period of coding sequence is longer than the wavelength of PW, which does not meet the criterion for PW-SW conversion under normal incidence. Note that this criterion was established for the ideal plane wave incident on an ideally gradient index metamaterial, whereas in our simulations,

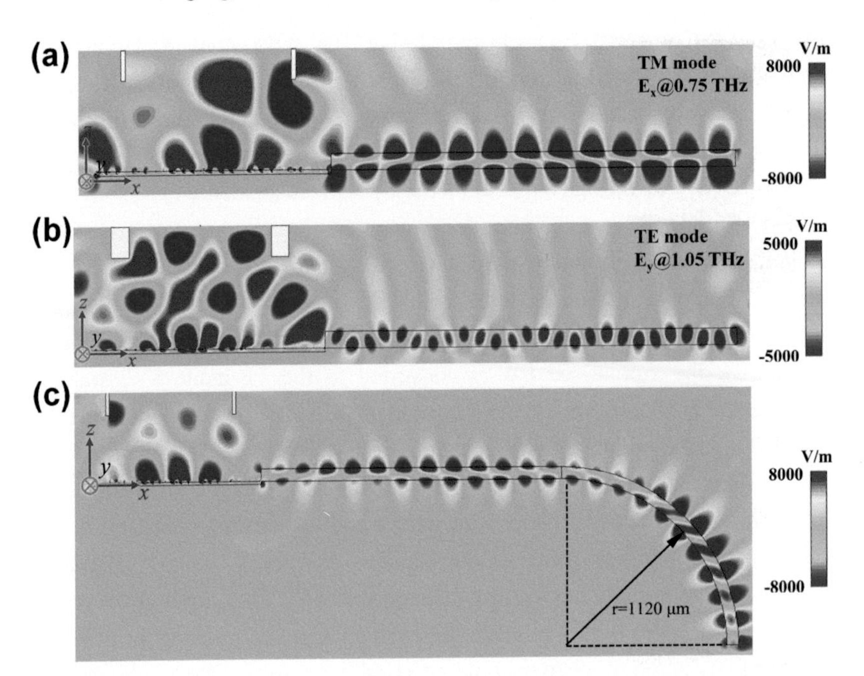

Figure 3.5 Simulation results of the PW-SW conversion of the tensor coding metamaterial. (a,b) Simulated electric field distributions (0.75 THz) of the TM-mode and TE-mode SWs, respectively. (c) Simulated electric field distribution (0.75 THz) of the TM-mode SW propagating on a curved quartz substrate.

neither of these requirements was met. As the beam generated by the waveguide port contains a wave vector in the x direction (propagation direction of SW), the momentum provided by the coding metamaterial can be reduced accordingly, which results in a larger period of the gradient coding sequence.

Figure 3.5b shows the simulated electric-field distribution of the TE-mode SW. In this case, because the coding sequence decreases to "1 3 5 7" with a smaller period, the optimum conversion effect appears at a higher frequency of 1.05 THz. To verify the correctness of the converted SWs, we carried out strict mathematical derivations and gave the analytical expression of the SWs in both TE and TM modes. Excellent agreements between the theoretical and numerical results verify that the wave observed on the quartz substrate is indeed SW. Due to the localization property of SW, the SW is tightly bounded to the curved quartz substrate with 1,120 μm curvature, as is shown in Figure 3.5c.

A THz time-domain near-field scanning system was employed to measure the SW propagating on the quartz substrate, as shown by the photograph presented in Figure 3.6a, which is briefly introduced as follows. A THz beam illuminates

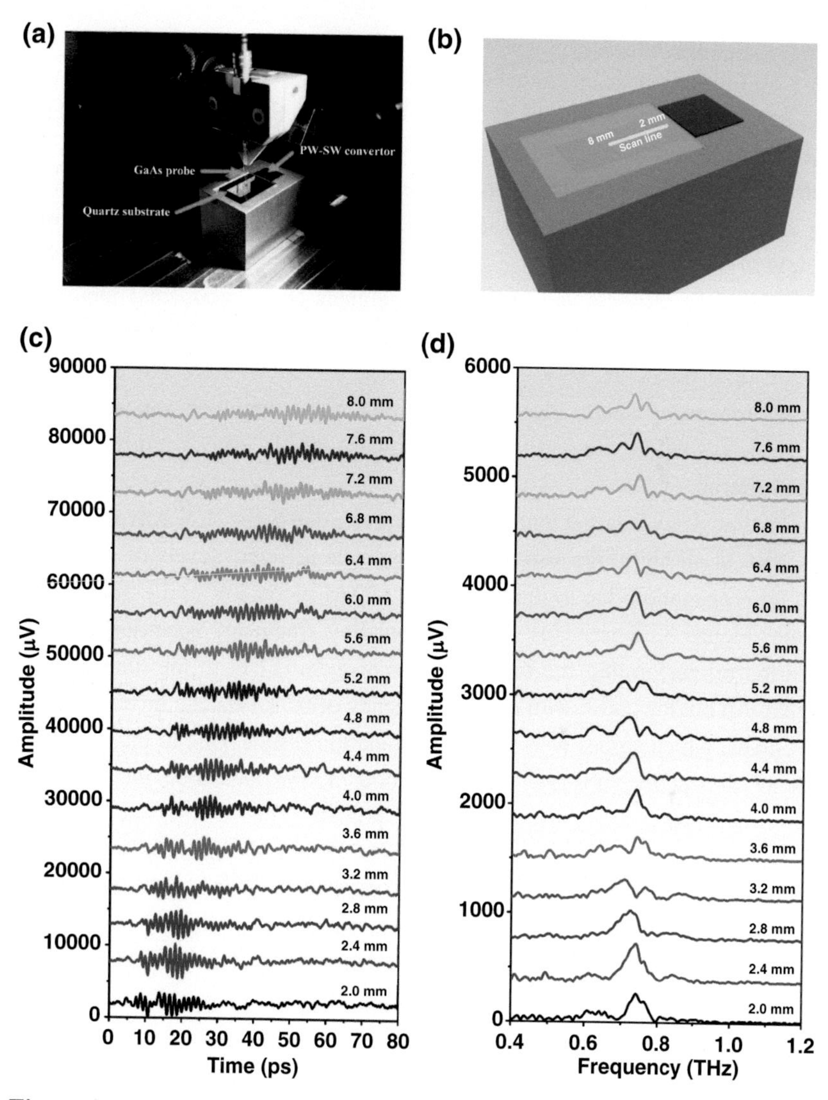

Figure 3.6 Experimental characterization of the tensor coding metamaterial at THz frequency. Schematic of the THz time-domain near-field scanning system. (b) Schematic of the sample holder. (c,d) Time-domain signals (Ez components), and frequency spectra of the SW measured every 0.4 mm from 2 to 8 mm on the middle line of the crystal quartz substrate.

the PW-SW convertor from underneath and is converted to SW that is guided to the 80 μm-thick crystal quartz substrate placed next to the coding metamaterial. Both the coding metamaterial and crystal quartz substrate are placed on a predesigned sample holder (see Figure 3.6b) to help align them with sufficient accuracy. A photoconductive antenna-based probe (Protemics GmbH), driven by a 2D translation stage, could automatically move in the x and y directions on the horizontal plane with high precision to measure the E_z components of the electric field 100 μm above the quartz substrate. Note that the THz wave reflected from the metal platform, where the sample holder is placed, cannot be received by the microprobe in the 80 ps time window due to the 20 mm height of the sample holder.

Figure 3.6c provides the time-domain signals (E_z components) measured every 0.4 mm from 2 to 8 mm on the middle line of the crystal quartz substrate in the time period of 0~80 ps, as indicated in Figure 3.6b. An oscillating waveform can be clearly identified from each curve, and their appearing time range delays as the measuring point gradually moves away from the coding metamaterial.

The time delay is larger than that in free space due to the slow wave nature of the SW. In addition, the waveforms experience different levels of distortion during propagation due to the dispersion of TM-mode SW in the quartz substrate. Figure 3.6d provides the corresponding frequency spectrum at each position by performing the fast Fourier transform (FFT) to the time-domain signal. An obvious peak with center frequency of 0.73 THz can be observed in all curves, which agrees well with the simulation results. Due to the dielectric loss of the quartz substrate, the amplitude of SW measured far from the convertor is lower than those at the nearer points.

4 Programmable Metamaterials

4.1 Digitally Controlled Coding Particles

4.1.1 1-Bit Digital Coding Particles

In the previous chapters, we have studied many types of coding metamaterials realized at various frequencies and with different types of waves. However, their functionality is fixed once they are fabricated. To allow dynamic control of the functionality of coding metamaterials, we need to design active coding particles whose digital states can be tuned as desired. The reflection/refraction phases of the coding particles can be switched electrically, mechanically, optically and thermally. We think that the ultimate and most competitive switching technique among all these approaches is electric control.

First, compared with the optical, mechanical and thermal approaches, the electric control does not need a big physical space because the feeding circuits can be embedded in the sample itself. This is not the case for the optical control method, as additional space is taken by bulky optical components that lie behind the coding particle.

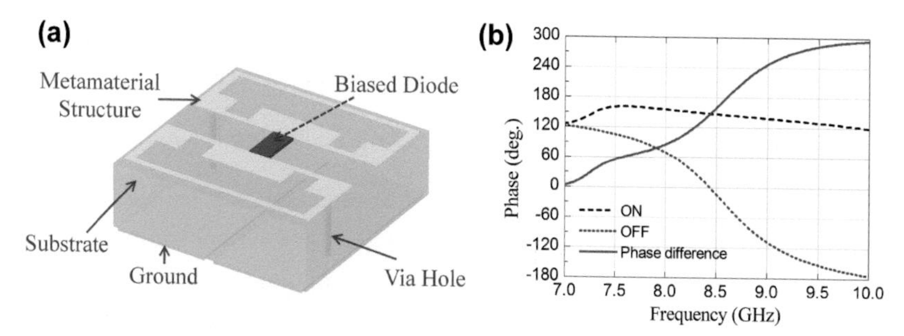

Figure 4.1 The first programmable metamaterial. (a) Structure of the digital particle. (b) Reflection phase curves of the digital particle in the ON and OFF states.

Second, the electric control method can provide much higher response speed than the thermal and mechanical control methods due to the fast speed of electron mobility. In addition, the thermal and mechanical control methods are vulnerable to environmental changes such as strong vibration and extreme temperatures. Owing to the solid state of all components involved in the electric control unit, it is more stable and reliable than the thermal and mechanical methods in practical applications.

Third, it is more convenient to combine the electrically controlled coding metamaterials with the digital circuits. For these reasons, we name the dynamically controlled coding particle as the digital particle.

As an example of practical implementation with realistic materials and dimensions, we designed and fabricated a digital coding metamaterial working at the microwave frequency. The constituent digital particle is illustrated in Figure 4.1a, which is similar to the sandwich design of most coding particles. The structure on the top of the dielectric spacer was made of F4B substrate ($\varepsilon_r = 2.65$ and loss tangent $\delta = 0.001$) and consisted of two hollow metallic strips. A pin-diode (Skyworks, SMP-1320) was welded between the two hollow strips and can be electrically switched ON and OFF through the feeding lines printed on the bottom side of the substrate.

Two metallic vias were drilled to connect the top hollow metallic strips with the bottom feeding lines. By setting the corresponding circuit parameters of the pin-diode in the CST Microwave Studio, we obtained the numerical results of the reflection phases when the pin-diode is in the ON and OFF states, as shown in Figure 4.1b. The phase difference approaches 180° from 8.3 to 8.9 GHz and reaches exactly 180° at 8.6 GHz.

We noted that the switching speed of the digital state determines the transmission rate when the coding metamaterials are used for communication applications. Hence, the pin-diodes with shorter recovery time are desired. The minimum recovery time reported for commercial pin-diodes was around 10 ns, which supports a maximum modulation speed of 100 MHz.

4.1.2 2-Bit Digital Coding Particles

Since there are only two states for a pin-diode, one can only realize the 1-bit digital particle using the structure described in Section 4.1.1 containing a single pin-diode. Apparently, two pin-diodes must be incorporated and independently biased in each unit cell to realize a 2-bit digital coding particle. Huang et al. recently developed a 2-bit digital particle by loading two independent pin-diodes into one unit cell, as illustrated in Figure 4.2a.

Two different dielectric substrates, 0.127 mm-thick substrate (Rogers 5880) and 3 mm-thick F4B substrate, were bounded together to form the dielectric spacer through a 0.09 mm-thick dielectric film (RO4403). On the top side of the 0.127 mm-thick substrate, two pairs of metallic trapezoidal patches were etched, where two pin-diodes (M/A-COM Flip Chip MA4SPS502) were welded between the two gaps. The back side of the F4B substrate was fully covered by metal, which was connected to the central trapezoidal patches to

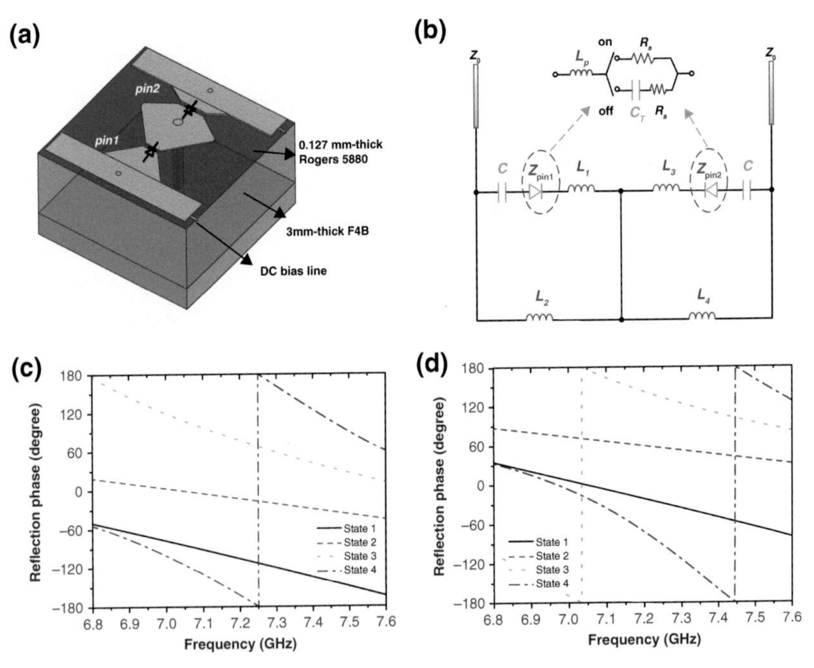

Figure 4.2 2-bit programmable metamaterial. (a,b) Structure and equivalent circuit model of the 2-bit digital particle. (c,d) Simulated reflection phases (7.25 GHz) of the four digital states “00”, “01”, “10” and “11” under the normal and 30° oblique incidences, respectively.

serve as the ground line. Two metal strips were connected to the trapezoidal patches to function as the positive DC bias line.

To obtain an analytical analysis of the reflection response of the 2-bit digital particle, we characterized it from the circuit perspective. Figure 4.2b shows the equivalent circuit model of the digital particle, in which the metallic parts and gaps between neighboring particles are modeled as inductors and capacitors, respectively. The pin-diode had different circuit models in the ON and OFF states. In the ON state, it was modeled as a series circuit of parasitic inductance and resistance, while in the OFF state, it was equivalent to a series circuit of parasitic inductance, capacitance and resistance. The effective impedance of the entire digital particle can be expressed as

$$Z_{eff}(\omega) = \frac{1}{\frac{1}{j\omega L_2} + \frac{1}{\frac{1}{j\omega C} + Z_{pin1} + j\omega L_1}} + \frac{1}{\frac{1}{j\omega L_4} + \frac{1}{\frac{1}{j\omega C} + Z_{pin2} + j\omega L_3}} \tag{4.1}$$

in which

$$Z_{pin} = \begin{cases} j\omega L_p + R_s & \text{"ON" States} \\ j\omega L_p + \dfrac{1}{j\omega C_T} R_s & \text{"OFF" states} \end{cases} \tag{4.2}$$

The circuit parameters of the pin-diode were experimentally obtained by applying a forward current If = 20 mA to switch the diode in the ON state, and a reverse bias voltage of 5 V for the OFF state. By fitting the measured data at the designed frequency 7.25 GHz, the circuit parameters were obtained as $R_s = 3.8$ ohm and $L_p =$ 0.65 nH in the ON state, and $C_T = 0.085$ pF and $L_p = 0.65$ nH in the OFF state. The reflection coefficient of the digital particle could then be expressed as

$$r(\omega) = \frac{Z_{eff}(\omega) - Z_0}{Z_{eff}(\omega) + Z_0} = |R|e^{j\omega\phi} \tag{4.3}$$

Figure 4.2c shows the simulated reflection phases of the four digital states "00", "01", "10" and "11", in which the adjacent states have a phase difference of 90° ±10° at the designed frequency 7.25 GHz. Due to the PEC ground sheet on the back of the substrate, the reflection amplitude was above 0.85 for all four digital states (see Figure 4.2d). The influence of feeding lines on the reflection/refraction properties of the digital particles was evaluated by simulating the structures with and without the bias lines, and no significant difference was found between these two scenarios.

In most cases, it is impractical to illuminate the coding metamaterials with an ideal plane wave at the normal incidence, as explained in Section 4.2. Hence, it is necessary to analyze the performance of the designed digital particle under

30° oblique incidence. Simulation results show that the phase difference between adjacent digital states becomes 90°±20°, which has larger tolerance than in the case of normal incidence. The reflection amplitude decreases slightly to 0.82, which is still acceptable in real applications.

4.2 Field-Programmable Metamaterials

4.2.1 Programmable Metamaterials under Plane Wave Excitation

In the previous chapters, we introduced the experimental realization of 1-bit and 2-bit digital particles. Similar to the coding metasurfaces, in which different coding particles are arranged in array to have certain manipulation to the EM waves, we can also arrange many of these digital particles in arrays and control the digital states of each digital particle independently. As the manipulations to the EM waves by such arrays can be dynamically controlled through the input coding sequences, we can call them programmable metasurfaces.

Different from the difficulties faced with the passive coding metasurfaces, one of the challenges in designing the programmable metasurface is the feeding circuit. For a programmable metasurface having $N\times N$ digital particles, a total number of N^2 feeds lines to control all particles independently, which may exceed the number of available output pins for a single microcontroller, such as an FPGA and Acorn RISC Machine (ARM). A smart feeding technique should be developed to reduce the required feeding lines. For instance, by adding shift register and latches to each unit cell, Ref. 182 presents a progressive scanning technique to control a 160×160 reflect array with only 162 control lines. Such a technique had been widely applied in the control of LC display.

As a proof of this principle, we present a simple demonstration of programmable metasurface in the microwave frequency, which consists of 30×30 digital particles. Considering the easy design of the layout of bias circuit, the digital particles were controlled column by column, with each column consisting of 30×5 digital particles. An FPGA was used to control the programmable metasurface by inputting six-digit coding sequences, as shown in Figure 4.3a. Three coding sequences demonstrate the programmable feature of the programmable metasurface. Figures 4.3b–d demonstrate the simulated and measured radiation patterns for the coding sequences “111111”, “010101” and “001011”, respectively. Both simulation and experimental results illustrate that the radiation patterns can be easily programmed by changing the corresponding coding sequences. We remark that, by elaborately designing the feeding circuit, we can realize 2D programmable metasurfaces, in which the digital particles are independently controlled to provide more freedom in manipulating the EM waves.

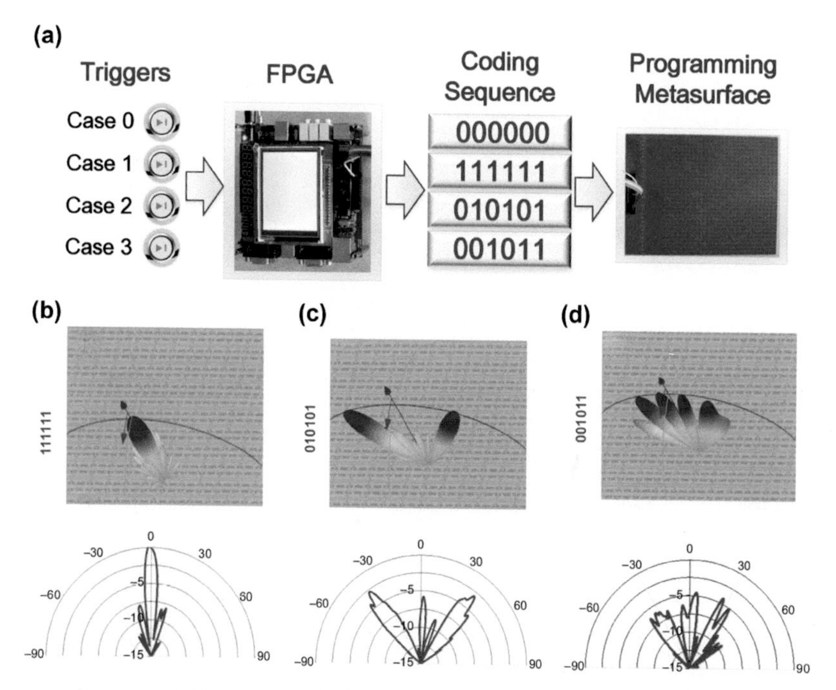

Figure 4.3 Experimental characterization of the first programmable metamaterial at microwave frequency. (a) Flow chart of the working mechanism of the 1-bit microwave programmable metamaterial. (b,c,d) Simulated and experimentally measured radiation patterns for the coding sequence "111111", "010101" and "001011", respectively.

4.2.2 Programmable Metamaterials under Point-Source Excitation

The previous subsection introduced the first programmable metasurface working under the normally incident plane wave. In real applications, the reflection-type programmable metasurfaces are commonly fed by a small antenna, which cannot provide ideal plane wave illumination. A quasi-plane wave assumption can be made only if the aperture of the feeding antenna is larger than the programmable metasurface, under which condition most of the radiated EM energies will be blocked.

To allow the programmable metamaterials to function normally under the excitation of small antenna, we proposed a strategy for designing the programmable metamaterials to overcome the nonplanar wavefront [183], as schematically illustrated in Figure 4.4a. For simplicity, the small feeding antenna was considered a point source that generates a spherical wavefront. In this scenario, the real phase reflected by each digital particle will deviate from the designed value. In order to maintain the desired phase distribution on the programmable

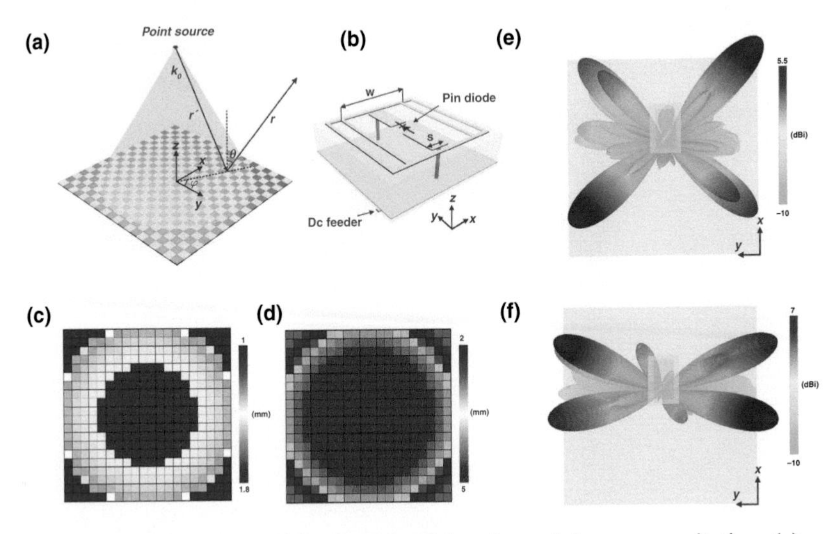

Figure 4.4 Programmable metamaterial under point source excitation. (a) Schematic illustration of the point source excited programmable metamaterial. (b) Structure of the digital particle. (c,d) Distributions of the optimized values of S and W on the 20×20 programmable metasurface, respectively. (e,f) Simulated radiation patterns of the chessboard and rectangular-type chessboard coding patterns under the illumination of rectangular feeding antenna.

metamaterial, the reflection phase of each digital coding particle has to be independently compensated, according to their distance relative to the point source, to retain their original value.

Before we describe the compensation method, it is necessary to give an analytical expression for the radiation pattern of the programmable metamaterial under the point source illumination. Due to the subwavelength nature of the digital particle, the detailed radiation characteristic of each coding particle becomes vague in the far-field region. Therefore, it is safe to assume an isotropic radiation pattern for each coding particle in calculating the radiation pattern of the entire programmable metamaterial:

$$F(\theta,\varphi)=\sum_{m=1}^{M}\sum_{n=1}^{N}\exp\{j[\varphi'(m,n)+kD_x(m-1/2)\sin\theta\cos\varphi + kD_y(n-1/2)\sin\theta\sin\varphi]\}, \tag{4.4}$$

in which $\varphi'(m,n)=\varphi(m,n)-kr'_{m,n}$. Clearly, the deviation of the phase profile comes from the different phase delays of the spherical wavefront illuminated at different positions on the programmable metamaterial. We attempted to maintain the desired phase distribution by changing the geometrical parameters the of digital particles, according to their relative distances to the point source. Since the relative position between the point source and metamaterial is fixed in real applications, the compensated programmable metamaterial should function correctly for any coding patterns.

A new structure shown in Figure 4.4b was designed in building the programmable metamaterial to verify the proposed method. A pin-diode with the same type as that adopted in Section 4.2 was welded at the center of the digital particle and biased with two feeding lines connected on the back layer through two metallic vias. Using the CST Microwave Studio, the optimized geometrical parameters can be obtained to realize the opposite reflection phases at the designed frequency of 8.9 GHz by switching the diode in the ON and OFF states.

By changing the geometrical parameters of the structure, S and W, the reflection phase will shift in a certain range for both the ON and OFF states. To verify the effectiveness of the proposed compensation technique, a programmable metamaterial with 20×20 digital particles was built, which was illuminated with a rectangular horn antenna placed at 250 mm above the metamaterial. Figures 4.4c and 4.4d demonstrate the distributions of the optimized values of S and W on the programmable metamaterial, respectively.

Due to the spherical wavefront of the point source, we could see that the distributions of the geometrical parameters S and W also exhibited the signature of circular rings, which could offset the spherical wavefront and allow the programmable metamaterial to function the same as that under the plane-wave illumination. Because the largest incident angle is lower than 20°, for the current height of the feeding antenna, it is safe to assume normal illumination for all digital particles in the programmable metamaterial. Figures 4.4e and 4.4f illustrate the simulated radiation patterns of the regular-type chessboard coding pattern and rectangular-type chessboard coding pattern under the illumination of the rectangular feeding antenna. Excellent agreement can be found by comparing them with the radiation pattern obtained under the plane wave illumination, which demonstrates the effectiveness of the compensation technique.

Experiments were conducted to further validate the performance of the designed programmable metamaterial under the point source illumination. Figure 4.5a shows the experimental configuration of the point-source-excited programmable metamaterial, in which a trestle was designed to keep the metamaterial and horn antenna in the right position. The programmable metamaterial consisted of 20×20 digital particles and was divided into 5×5 independent control units (see Figure 4.5b), with each control unit independently controlled by FPGA.

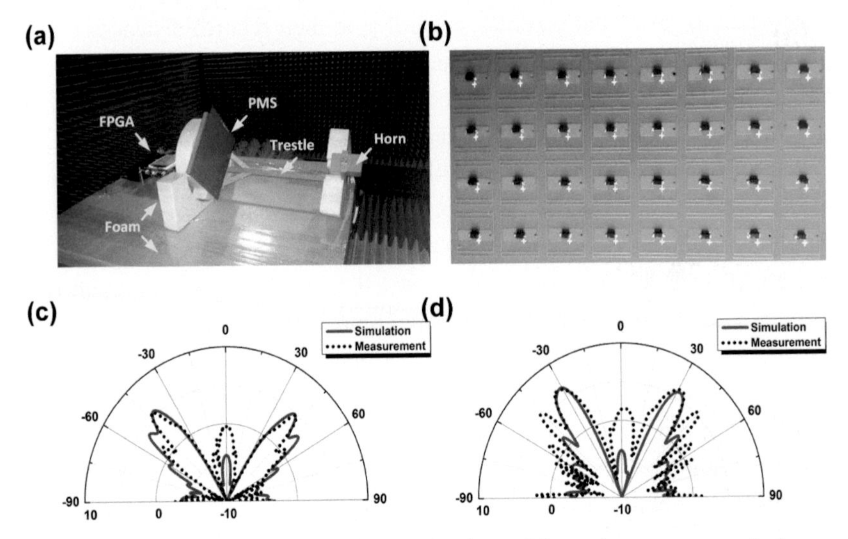

Figure 4.5 Experimental characterization of the point-source-excited programmable metamaterial at microwave frequency. (a) Experimental configuration. (b) Fabricated sample. (c,d) Measured radiation patterns of the chessboard and rectangular-type chessboard coding patterns, respectively.

Figures 4.5c and 4.5d show the measured radiation patterns of the regular-type chessboard and rectangular-type chessboard coding patterns, which were obtained by simply changing the input coding sequence stored in FPGA. The good match between simulated and measured results further validates the effectiveness of the compensation method.

The programmable metamaterials were further extended to large scale recently, which incorporates 40×40 digital particles, as shown in Figure 4.6a. [184] It is made of five identical sub-metamaterials, each including 8×40 digital particles. The entire programmable metamaterial has an electrical length of more than 20 free-space wavelengths. To reduce the complexity of the feeding circuit and control system, 200 shift registers were adopted, each of which controls eight pin-diodes in a sequential manner. The top view of the digital particle is shown in Figure 4.6b, in which the pin-diode (MACOM MADP-000907–14020) is welded between a square patch and a metallic via. To choke the RF signal and increase the isolation between the DC and RF responses, a quarter-wavelength micro-strip line and an open-ended radial stub were added to the lower part of the structure. [185] The back side of the dielectric substrate (Taconic TLX-8) was fully covered by a metal sheet.

Using the CST Microwave Studio, the reflection phases under the x and y polarized incidences were obtained, which showed 180° and 0° phases when

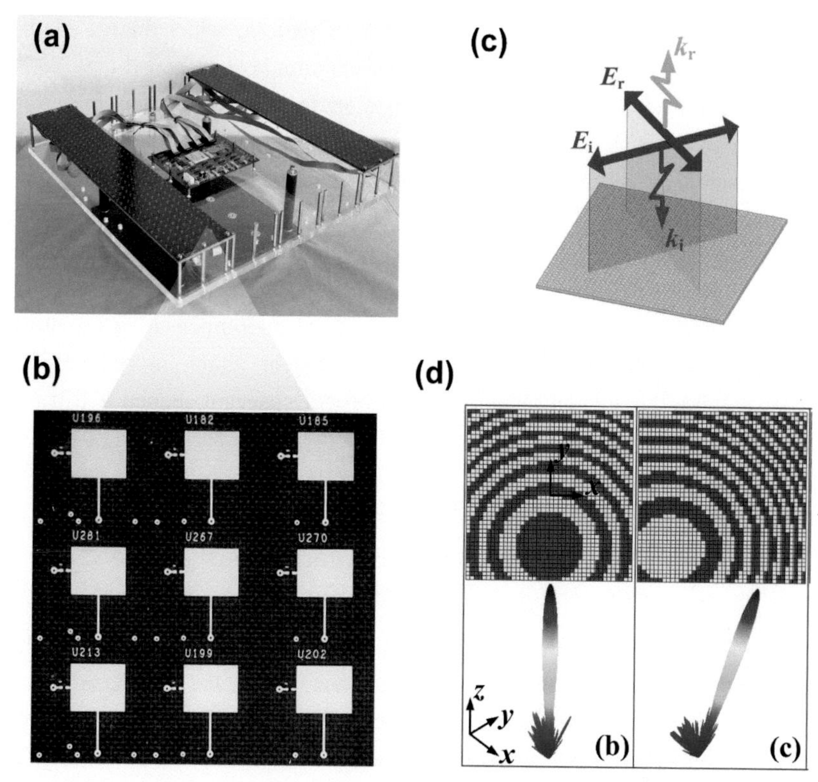

Figure 4.6 Large-scale programmable metamaterial. (a) Prototype of the large-scale programmable metamaterial having 40×40 digital particles. (b) Enlarged view of the structure of the coding particle. (c) Illustration of the functionality of 90° linear polarization conversion. (d) Simulated radiation patterns of two coding patterns.

the diode was biased ON and OFF, respectively. The phase difference between the x and y polarizations comes from the geometrical anisotropy of the structure, and can be used to realize the function of 90° linear polarization conversion when the incident wave is polarized at 45° with respect to the x/y axis, as illustrated in Figure 4.6c. Interestingly, such a functionality of polarization conversion can be dynamically switched on and off by controlling the applied voltages. For example, when all the diodes are biased in the ON state, the 45° polarized linear incidence will be reflected back with the original polarization.

To make the programmable metamaterial work normally under the illumination of a small horn antenna, which was placed at 198 mm above the programmable metamaterial with an offset angle of 20° (with respect to the z axis), the genetic algorithm (GA) was employed to optimize the required coding patterns. To reduce the computational complexity and time, a 2D inverse Fourier transform was used to calculate the radiation pattern, which increased the optimization speed by at least two orders of magnitude.

Different from the previous compensation technique, in this work, the spatial phase delay that resulted from the point source excitation was compensated by discretizing the continuous compensation phase into binary phases with the "1" and "0" states. Figure 4.6d illustrates two coding patterns and their corresponding radiation patterns, which were designed to generate a single-beam radiation pointing in the 0° and 20° directions. To demonstrate the effectiveness of the GA method in compensating the point source illumination with certain oblique incident angle, more complex beam patterns have been tested, such as broad, cosecant-shaped and triple beams.

5 Information Processing on Digital Coding Metamaterials

5.1 Relationship between Digital Coding Pattern and Far-Field Response

It is well known that radiation in the far-field region can be expressed in the form of the Fourier transform of the near electric field on the antenna aperture. Based on this relationship, we can easily obtain the far-field radiation pattern of a digital coding metasurface without numerical simulations, which commonly take considerable time and computational resources. In this section, we briefly introduce the process for calculating the radiation pattern from a given coding pattern. It mainly involves the Fourier transform of the coding pattern, and a subsequent coordinate transformation from the angular coordinate to the visible angle coordinate, as illustrated in Figure 5.1. The first step is to calculate the 2D FFT image of the coding pattern in Figure 5.1a, which gives the result in Figure 5.1b. Since the FFT image is expressed in the angular coordinate (u,v), it should be transformed to the visible angle coordinate using the following equations:

$$u = \frac{2\pi}{\lambda} d_x \sin\theta \cos\varphi \tag{5.1}$$

$$v = \frac{2\pi}{\lambda} d_y \sin\theta \sin\varphi \tag{5.2}$$

in which λ and d_x are the free-space wavelength and the period of coding element, respectively, and θ and φ are the elevation and azimuthal angles in the spherical coordinate system.

As (u,v) range from $-\pi$ to π, the coordinate transformation from (u, v) to (θ, φ) is not a bijection (one-to-one correspondence) and is dependent on the electrical

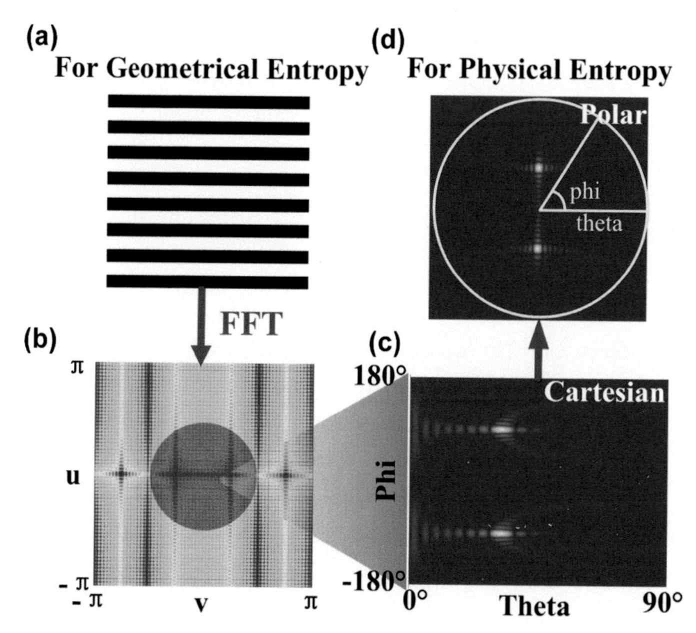

Figure 5.1 The process for calculating the radiation pattern from a given coding pattern. (a) Coding pattern. (b) The FFT image of the coding pattern. (c,d) The far-field radiation patterns expressed in the Cartesian coordinate and polar coordinate, respectively.

size of the coding particle. Only the data inside the red circle in Figure 5.1b are transformed to the visible angle coordinate. Apparently, three cases should be considered, which are $d = \lambda/2$, $d < \lambda/2$ and $d > \lambda/2$. For the first case, the circle is tangent to the square, in which condition we can just obtain the transformed visible angles at 90°. For the second case, the circle is smaller than the square, thus not all the data in the angular coordinate should be transformed to the visible angle coordinate. For the third case, the circle is larger than the square, hence some data at larger angles will be discarded after the coordinate transformation. Based on these analyses, it is concluded that the size of the coding particle should not be larger than half a wavelength so as to allow omnidirectional radiation.

Figure 5.1c is the result of radiation pattern after the coordinate transformation, in which the horizontal and vertical axes represent the elevation and azimuthal angles, respectively. To have a clearer view of the radiation pattern, it is further transformed into the polar coordinate in Figure 5.1d, in which the radial and axial directions represent the θ and φ coordinates, respectively. Note that only the data inside the circle are valid. As seen from the following examples, the excellent agreement between the numerical simulations and theoretical calculations validates the accuracy of the FFT method, which serves as an efficient tool in calculating the radiation patterns of the electrically large coding patterns at a speed 3~4 orders of magnitude faster than the full-wave simulations.

5.2 Information Entropy of Digital Coding Metamaterial

Entropy was initially proposed to describe the macroscopic behaviors of a thermodynamic system that consists of a large number of constituents (atoms or molecules). The concept of entropy was later extended to other fields like the von Neumann entropy in quantum mechanics, and information entropy (also referred to as Shannon entropy) in the field of information theory. Shannon entropy, proposed by Calude Shannon in 1948, provides an absolute limit on the best possible average length of loss-less encoding or compression of an information source [186]. The larger the entropy is, the more random the information source will be, and the more information it carries.

As is known from previous sections, the programmable metasurface has the capability to generate many different radiation patterns dynamically. In this regard, the programmable metasurface could be considered as an information source, in which the information is modulated on the radiation patterns in the far field.

Figure 5.2a sketches the conceived wireless communication system based on the programmable metasurface. Similar to the conventional wireless communication system, it involves three major parts: the transmitter, receiver, and channel. Multiple receivers need to be deployed at different positions in the far-field region to receive the transmitted signals after they pass through the wireless channel with multipath effect and noise interference. It would be interesting, and also of significant importance, if we could estimate the amount of information carried by a certain coding pattern.

Based on the definition of information entropy, we proposed, for the first time, a fast method to calculate the information entropy of a coding pattern using the following function [187]:

$$H_2 = -\sum_{i=1}^{2}\sum_{j=1}^{2} P_{ij} log_2 P_{ij} \tag{5.3}$$

in which P_{ij} is the joint probability of a group *G(i, j)* used to indicate two adjacent coding elements, as illustrated in Figure 5.2b. As this is the 2D information entropy, it reflects not only the proportion of the "1" and "0" elements in the coding pattern but also their spatial distributions. As apparent from Equation 5.3, the appearance of the four groups *G(0, 0)*, *G(0, 1)*, *G(1, 0)*

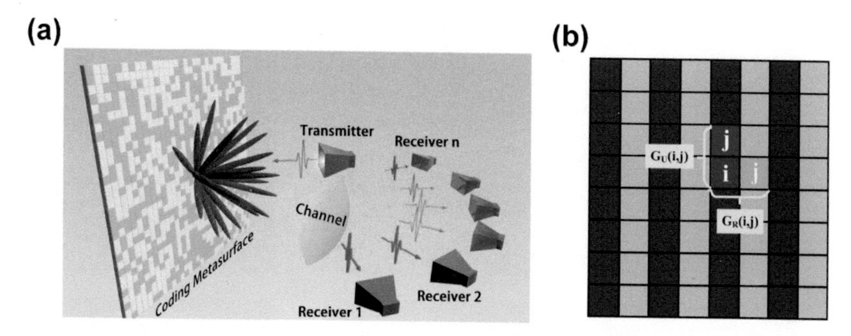

Figure 5.2 Information entropy of coding metamaterials. (a) Schematic illustration of a programmable metamaterial-based wireless communication system. (b) Method for the calculation of information entropy of a coding pattern.

and $G(1, 1)$ determines the 2D entropy. As the group $G(i, j)$ can be on either the right side ($G_R(i, j)$) or the upper side ($G_U(i, j)$) of the current code (indicated by the red color), the final entropy should be the average value of the entropies calculated in row (H_{2R}) and column (H_{2U}).

To find the relationship between the entropies of the coding pattern (geometrical entropy) and radiation pattern (physical entropy), different coding patterns were tested, including periodic and nonperiodic patterns. As the electrical size of the coding particle is only $\lambda/5$, the radiation pattern can cover the entire upper-half space. Note that the radiation pattern should be expressed in the polar coordinate and transformed to gray color before it can be used for the calculation of physical entropy.

Figures 5.3a–c give three different coding patterns generated by a cellular automata machine, in which a half "1" and half "0" coding pattern (Figure 5.3a) is gradually diffused into a series of random coding patterns by exchanging two adjacent coding particles randomly in each time. Figures 5.3b and 5.3c present the coding patterns selected at the middle and final stages of this diffusion process. Using the FFT method, we obtained their radiation patterns, shown in Figures 5.3d–f. The geometrical entropies of the three coding patterns increases from left to right, which indicates the randomness of the distribution of "0" and "1" coding particles. The physical entropies of their radiation patterns also increase accordingly, due to the increasing random scattering beams.

To further verify the relationship between the geometrical entropy and the physical entropy, we provide in Figure 5.3g the entropies of the 99-iteration coding patterns and radiation patterns obtained in the diffusion process. It can be observed that the physical entropy increases roughly with the increase of geometrical

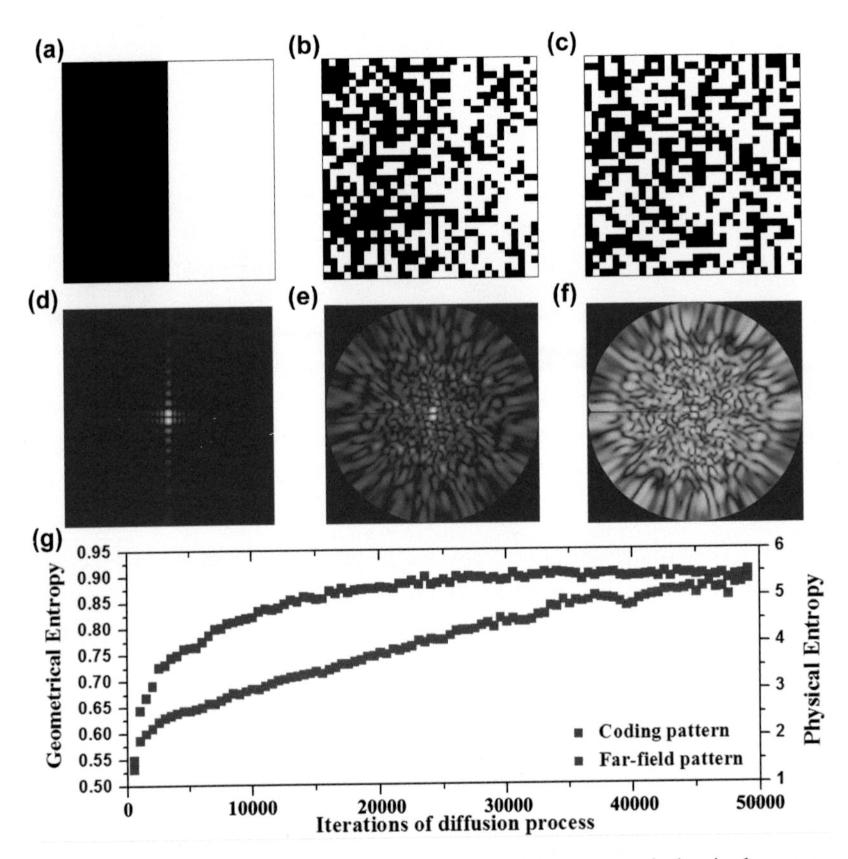

Figure 5.3 Simulation results of the geometrical entropy and physical entropy of coding metamaterials. (a–c) Three different coding patterns with increasing randomness. (d–f) The calculated radiation patterns corresponding to (a–c), respectively. (g) Geometrical entropy and physical entropy of the 99 coding patterns obtained in the diffusion process.

entropy. Interestingly, this process mimics the mixing of two types of gas molecules in a container, in which the system entropy also experiences an increasing trend. With the proportional relationship between the physical entropy and geometrical entropy, we can control the information radiated by the programmable metasurface by generating coding patterns with desired geometrical entropy, which are potentially useful in designing new functional devices and information systems.

5.3 Convolution Operations on Digital Coding Metamaterials

As is known from the previous chapters, the anomalous reflection/refraction angle of a coding metasurface is determined by the period of the gradient coding sequence. For the previous coding scheme, the minimum gradient coding sequence, for instance, "0 1 2 3 . . . " (super unit cell size of 1×1), is multiplied by an integer. As a consequence, the attainable reflection/refraction angle will be limited to certain discrete values, seriously restricting the application scope of the digital coding and programmable metasurfaces.

To enable continuous beam scanning for the coding metasurfaces, we proposed a new coding strategy, called the scattering pattern shift, which can rotate a radiation pattern to an arbitrary direction with negligible distortion [156] by applying the convolution theory from the digital signal processing to the coding metasurface. With the coding strategy, we could easily generate many complicated radiation patterns, which commonly requires brutal-force numerical simulations based on the previous coding scheme.

Before we introduce the principle of the scattering pattern shift, we should reiterate that the far-field radiation pattern and the coding pattern are a Fourier transform pair. Based on the Fourier transform relation, we could possibly apply some existing theorems in the Fourier transform, for instance, the convolution theorem, to design the coding metasurface, to enable more flexible manipulations of the EM waves. Here, the basic principle of the convolution theorem is briefly reviewed with the following equation:

$$f(t) \cdot g(t) \xleftrightarrow{FFT} f(\omega) * \ g(\omega) \tag{5.4}$$

This equation describes the equivalence between the ordinary multiplication of two time-domain signals and the convolution of their corresponding frequency spectra in the frequency domain. But the unity amplitude will no longer keep after the convolution operation. To preserve the correct digital state for the coding metasurface, which is to keep their unity amplitude after the convolution operation, a simplified version of the convolution theorem should be considered by assuming the item g(ω) as a Dirac-delta function,

$$f(t) \cdot e^{j\omega_0 t} \xleftrightarrow{FFT} f(\omega) * \delta(\omega - \omega_0) = f(\omega - \omega_0) \tag{5.5}$$

in which $e^{j\omega_0 t}$ is the time-shift item in the time domain and the impulse function $\delta(\omega - \omega_0)$ is its frequency spectrum expression in the frequency domain. Equation (5.5), also known as the frequency shift theorem, indicates that the convolution of a spectrum $f(\omega)$ with an impulse function $\delta(\omega - \omega_0)$ results in a shift of the spectrum function $f(\omega)$ by a value of ω_0 in the frequency domain without distortion.

Thus an interesting intuition arises: we should be able to apply Equation (5.5) to the coding metasurface if the variation of the coding and radiation patterns are viewed as the time-domain signal and frequency-domain signal, respectively.

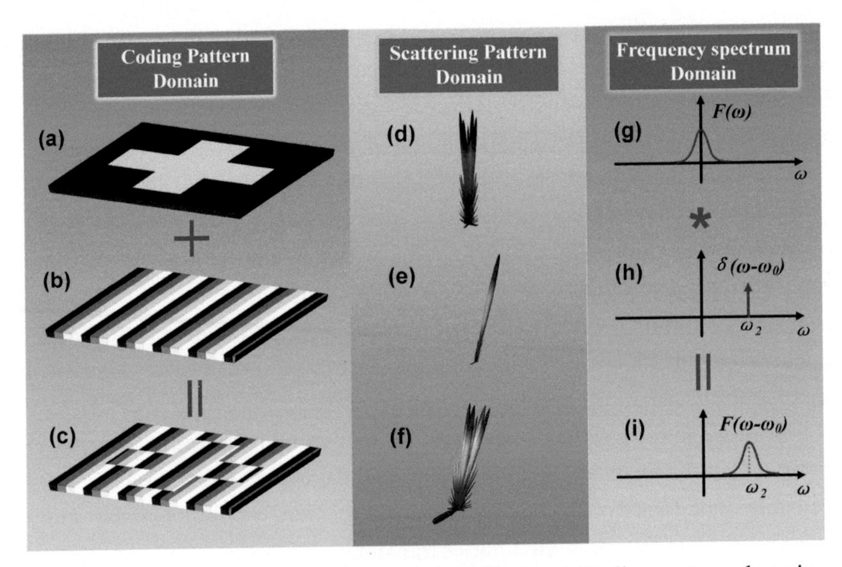

Figure 5.4 Principle of scattering pattern shift. (a–c) Coding pattern domain. (d–f) Calculated radiation patterns corresponding the coding patterns in (a–c), respectively. (g–i) The equivalent frequency spectra of the radiation patterns in (d–f), respectively.

This intuition is correct, as clearly demonstrated by the following variable substitutions, where t and ω are replaced by x_λ and $sin\theta$, respectively, giving the function of the scattering pattern shift as

$$E(x_\lambda) \cdot e^{jx_\lambda \sin\theta_0} \xleftrightarrow{FFT} E(\sin\theta) * \delta(\sin\theta - \sin\theta_0) = E(\sin\theta - \sin\theta_0) \tag{5.6}$$

in which the item $e^{jx_\lambda \sin\theta_0}$ represents the coding pattern with gradient phase along a certain direction. Similar to the frequency shift theorem, Equation (5.6) describes the equivalence between the multiplication of a coding pattern $E(x_\lambda)$ with a gradient coding pattern $e^{jx_\lambda \sin\theta_0}$ and the convolution of their radiation patterns. The resulting radiation pattern $E(\sin\theta - \sin\theta_0)$, inheriting most of the features of the original coding pattern, is diverted to the main beam direction of the gradient coding pattern. Note that the new radiation pattern experiences certain distortion during the rotation process, but can be neglected in most cases for small rotation angles.

To have a clear understanding of the concept of the scattering pattern shift, three different coding patterns based on 2-bit coding metasurface are taken as examples, as shown in Figures 5.4a–c. A cross-shaped coding pattern has a five-beam radiation pattern (see Figure 5.4d), which can be rotated to an anomalous direction (see Figure 5.4f) without observable distortion by making a convolution operation with the single-beam radiation pattern (see Figure 5.4e). Such a rotation can be accomplished by multiplying the cross-shaped coding pattern (see Figure 5.4a) with the gradient coding pattern "0 1 2 3 . . . " (see Figure 5.4b) in the coding pattern domain.

We note that the multiplication of two coding patterns is simply the modulus of their coding digits, resulting in the mixed coding pattern in Figure 5.4c, which inherits both features of the constituent coding patterns in Figures 5.4a and 5.4b. It is interesting to find that the rotation process mimics the frequency-shift theorem, as schematically illustrated in Figures 5.4g–i, where the original frequency spectra (see Figure 5.4g) is shifted to a higher frequency ω_2 (see Figure 5.4i) after being made the convolution operation with the Delta function in Figure 5.4h. We remark that with this new coding strategy, one can easily generate, at negligible computational complexity, various complicated radiation patterns that commonly require brutal-force simulations under the conventional coding scheme.

Figure 5.5 provides three different examples to demonstrate the performance of the principle of scattering pattern shift. All four coding patterns have the same number of coding particles of 64×64. The first coding pattern "0 2 0 2 . . . " with super unit cell size of 8×8 (see Figure 5.5a) generates a two-beam radiation pattern in the *x-z* plane (see Figure 5.5e). By adding a gradient coding sequence "0 1 2 3 . . . " with super unit cell size of 3×3 to the original coding pattern, we obtain the mixed coding pattern in Figure 5.5b, which generates a two-beam radiation pattern that is tilted away from the *x-z* plane (see Figure 5.5f). The deviation angle of 20.9° is exactly the same as the anomalous reflection angle of the gradient coding pattern.

The second and third examples are given to demonstrate the ability of the new coding strategy to generate single-beam radiation pointing in an arbitrary direction. Two gradient coding sequences with different super unit cell sizes, "0 0 1 1 2 2 3 3 . . . " and "0 0 0 1 1 1 2 2 2 3 3 3 . . . ", are added and subtracted, resulting in the mixed coding patterns in Figures 5.5c and 5.5d, respectively. It is interesting to notice that, when the two coding sequences are added, the variation of code is faster than the constituent coding sequences (see Figure 5.5c), leading to a large anomalous radiation angle of 63.2° (see Figure 5.5g); if they are subtracted from each other, the coding pattern exhibits a slower variation rate (see Figure 5.5d) and results in a small anomalous radiation angle of 10.3° (see Figure 5.5h). Note that the reflection angle θ of the mixed coding pattern is not the direct sum or subtraction of the radiation angles θ_1 and θ_2 of the constituent coding sequences, but should be calculated using the following function:

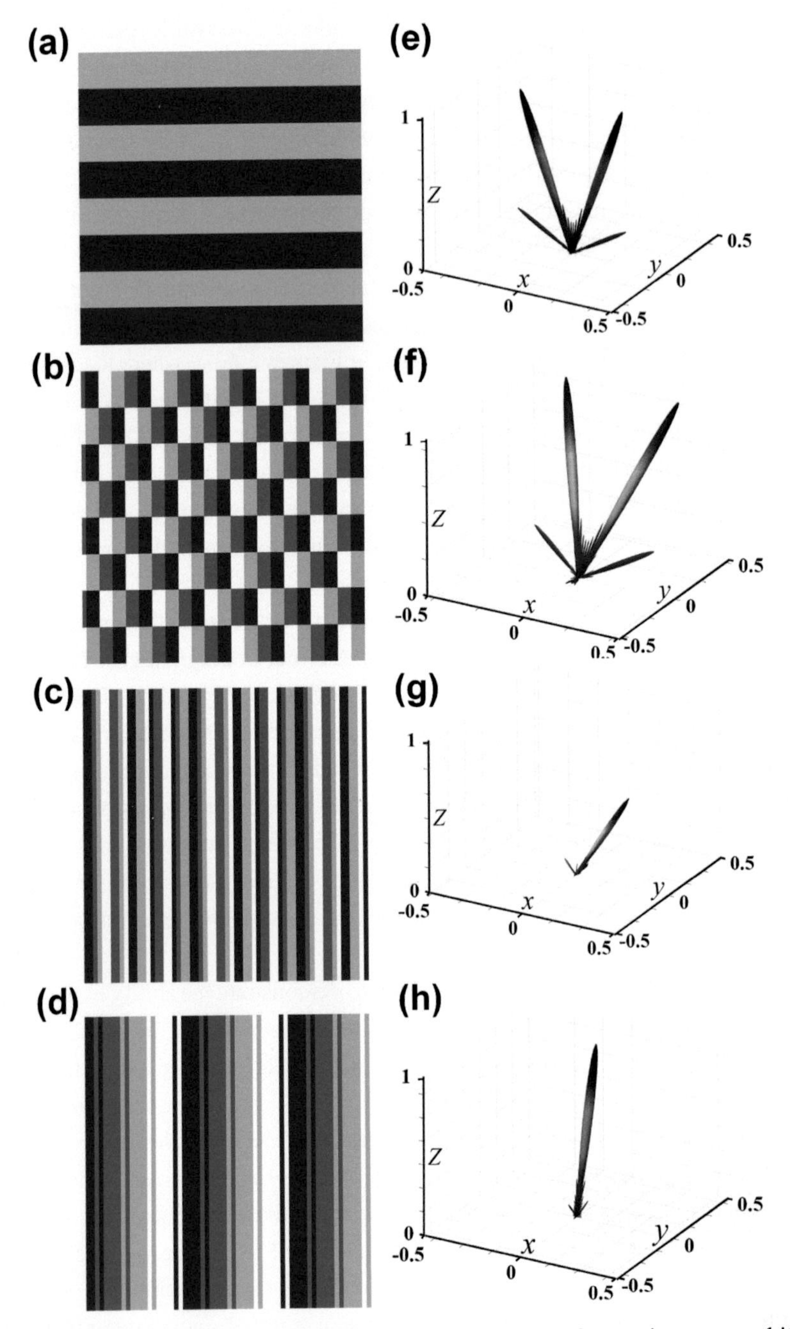

Figure 5.5 Perfect rotation effect enabled by the principle of scattering pattern shift. (a) "0 2 0 2 . . . " sequence with super unit cell of 8×8. (b) The addition of the coding pattern in (a) and a "0 1 2 3 . . . " sequence with super unit cell size of 3×3. (c) The addition of "0 0 1 1 2 2 3 3 . . . " and "0 0 0 1 1 1 2 2 2 3 3 3 . . . " sequences. (d) The subtraction of "0 0 1 1 2 2 3 3 . . . " and "0 0 0 1 1 1 2 2 2 3 3 3 . . . " sequences.

$$\theta = \sin^{-1}(\sin\theta_1 + \sin\theta_2) \tag{5.7}$$

It is obvious from Equation (5.7) that for small angles θ_1 and θ_2, the radiation angle θ of the mixed coding pattern is approximately the sum of θ_1 and θ_2. Little distortion is introduced to the new radiation pattern during this convolution operation process. However, for larger θ_1 and θ_2, the distortion should be considered in some applications that require accurate radiation patterns. Note that a rotation without distortion could be achieved by adding a compensation coding pattern to the mixed coding pattern, which is our future research direction.

To demonstrate the continuous scanning ability of the principle of scattering pattern shift, Figure 5.6a shows the attainable radiation angles calculated from the addition of two gradient coding sequences with different super unit cell sizes M and N (i.e. repetition number). Note that in this case, both sequences vary along the same direction. The minus sign indicates the case when the sequence varies along the reverse direction. From the calculated result, it is clear that almost all values from 0° to 90° can be obtained. The angle is sparsely distributed at larger angle ranges compared to the smaller angle ranges, but can be improved by further increasing the super unit cell sizes M and N.

Similarly, we could also rotate the single-beam pattern in the azimuthal direction by adding a gradient coding sequence that varies along the orthogonal direction. The radiation angle of the single beam can be calculated by the following functions:

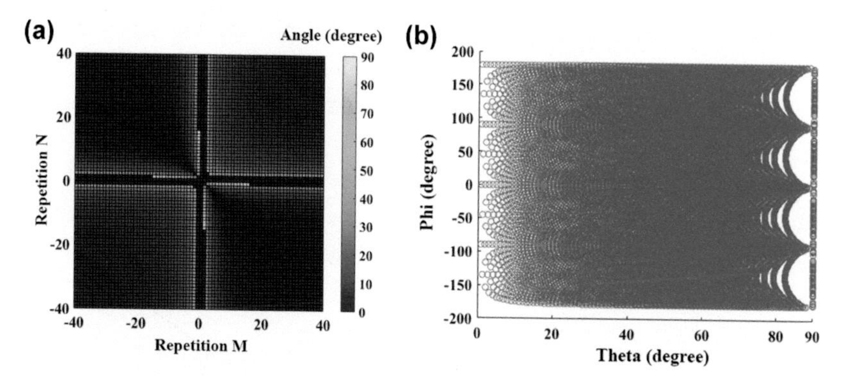

Figure 5.6 Demonstration of the continuous scanning ability of the scattering pattern shift coding strategy. (a) The attainable radiation angles calculated by adding two gradient coding sequences with different super unit cell sizes M and N. (b) The attainable angle of the mixed coding pattern when the two gradient coding sequences vary along orthogonal directions.

$$\begin{cases} \theta = \sin^{-1}\left(\sqrt{\sin^2\theta_1 + \sin^2\theta_2}\right) \\ \varphi = \tan^{-1}\left(\dfrac{\sin\theta_2}{\sin\theta_1}\right) \end{cases} \tag{5.8}$$

Figure 5.6b illustrates the result of the attainable scanning angle of the mixed coding pattern when the two gradient coding sequences vary along orthogonal directions. In the calculation, the radiation angle of the two gradient coding sequences is assumed to sweep from −90° to +90° with steps of 1°. It is clearly observed from Figure 5.6b, in which each circle represents a calculated result, that the single-beam radiation can cover the entire upper-half space with small steps. The four gaps near $\theta = 90°$ can be further covered by decreasing the angle step of the constituent gradient coding sequences.

5.4 Complex Coding and Addition Theorem

In parallel with the principle of scattering pattern shift, we proposed another coding scheme, called the addition theorem, which allows us to combine two independent coding patterns into one radiation pattern [188]. We notice that the phase response of a unit cell can be expressed as a complex form $e^{j\varphi}$ instead of φ itself, based on Maxwell's equations and wave equations. Then a complex code containing the full phase information of the EM waves was proposed, which possesses higher degrees of freedom in the coding pattern design to generate more flexible radiation patterns.

5.4.1 Addition Theorem for Complex Coding Schemes

Before we introduce the addition theorem for complex coding schemes, we should give the definition of complex digital codes. As shown in Figures 5.7a and 5.7b, all complex digital codes are located on a unit circle, which is named the "coding circle," on which an arbitrary-bit digital state is denoted by unit vectors on the coding circle. For simplicity, we list the abbreviations of the complex digital codes in Table 5.1, where the main body and subscripts represent the absolute value and bit number of the digital state, respectively.

The addition operation of two complex digital codes leads to:

$$e^{j\varphi_1} + e^{j\varphi_2} = Ae^{j\varphi_0} \tag{5.9}$$

which produces a new complex number with phase φ_0 and magnitude A. In fact, the addition operation can be realized by the vector superposition principle in traditional Euclidean geometry. For example, the addition of two 2-bit complex-codes $\dot{0}_2 + \dot{1}_2$ gives a phase of 45°, which corresponds to the complex digital state $\dot{1}_3$ in the 3-bit coding set, as illustrated in Figure 5.7a, while $\dot{0}_2+$ results in

Table 5.1 Abbreviations of complex digital codes

The Scale of Codes	Corresponding Digital States	Abbreviations of Digital States
1-bit Complex Codes	$\dot{0}$ $\dot{1}$	$\dot{0}_1$ $\dot{1}_1$
2-bit Complex Codes	$0\dot{0}$ $0\dot{1}$ $1\dot{0}$ $1\dot{1}$	$\dot{0}_2$ $\dot{1}_2$ $\dot{2}_2$ $\dot{3}_2$
3-bit Complex Codes	$0\dot{0}0$ $0\dot{0}1$ $0\dot{1}0$ $0\dot{1}1$ $1\dot{0}0$ $1\dot{0}1$ $1\dot{1}0$ $1\dot{1}1$	$\dot{0}_3$ $\dot{1}_3$ $\dot{2}_3$ $\dot{3}_3$ $\dot{4}_3$ $\dot{5}_3$ $\dot{6}_3$ $\dot{7}_3$

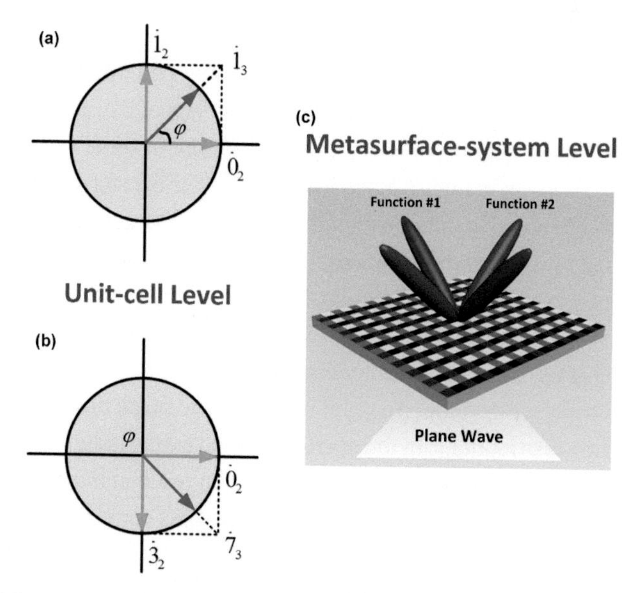

Figure 5.7 Intuitive demonstration of the addition theorem on the unit-cell and metasurface-system levels. Two typical addition processes of 2-bit complex codes (a) $\dot{0}_2 + \dot{1}_2 = \dot{1}_3$ and (b) $\dot{0}_2 + \dot{3}_2 = \dot{7}_3$ are shown on the coding circle. (c) Illustration of the system-level performance of the addition theorem.

a phase of 315°, corresponding to the complex digital state $\dot{7}_3$ in 3-bit coding, as illustrated in Figure 5.7b. These examples show that the addition of two n-bit complex codes produces an (n+1)-bit complex code. This means that we can obtain all higher-bit complex coding sets by iterating the addition operations several times from the simplest 1-bit complex codes.

However, there is a situation when the addition of two complex codes with opposite phases results in zero value, which is called an "indefinite coding

Table 5.2 The final regulations for addition operations on 1-bit and 2-bit complex digital codes

1-bit	$\dot{0}_1 + \dot{0}_1 = \dot{0}_2$		$\dot{0}_1 + \dot{1}_1 = \dot{1}_2$	
	$\dot{1}_1 + \dot{0}_1 = \dot{3}_2$		$\dot{1}_1 + \dot{1}_1 = \dot{2}_2$	
2-bit	$\dot{0}_2 + \dot{0}_2 = \dot{0}_3$	$\dot{1}_2 + \dot{0}_2 = \dot{1}_3$	$\dot{2}_2 + \dot{0}_2 = \dot{6}_3$	$\dot{3}_2 + \dot{0}_2 = \dot{7}_3$
	$\dot{0}_2 + \dot{1}_2 = \dot{1}_3$	$\dot{1}_2 + \dot{1}_2 = \dot{2}_3$	$\dot{2}_2 + \dot{1}_2 = \dot{3}_3$	$\dot{3}_2 + \dot{1}_2 = \dot{0}_3$
	$\dot{0}_2 + \dot{2}_2 = \dot{2}_3$	$\dot{1}_2 + \dot{2}_2 = \dot{3}_3$	$\dot{2}_2 + \dot{2}_2 = \dot{4}_3$	$\dot{3}_2 + \dot{2}_2 = \dot{5}_3$
	$\dot{0}_2 + \dot{3}_2 = \dot{7}_3$	$\dot{1}_2 + \dot{3}_2 = \dot{4}_3$	$\dot{2}_2 + \dot{3}_2 = \dot{5}_3$	$\dot{3}_2 + \dot{3}_2 = \dot{6}_3$

addition" because the zero value can be taken as any phase. This problem should be properly solved before we set rules for the addition operations. Based on theoretical analysis, we find that the indefinite-element rate of N-bit complex codes is estimated as $2^N/2^{2N} = 1/2^N$ for any coding pattern. Therefore, 1-bit complex codes suffer the biggest indefinite-element rate of 0.5, and will become even larger if the probability of every code is different in some special cases.

An example is given in Figure 5.8 to illustrate the performance of the addition theorem in combining two dual-beam radiation patterns (Figures 5.8a and 5.8b) into one quad-beam radiation pattern (Figure 5.8c). However, an obvious radiation is observed in the 0° direction from the resulting radiation pattern in Figure 5.8c, which is due to the missing digital states resulting from the massive indefinite elements in the additive coding metasurface.

To fix this issue, we developed a set of regulations to break this consistency of identical coding distributions on the aperture caused by the indefinite additions, while keeping the rationality. We chose the addition result at the bisector of the angle from the first complex code to the second complex code. For example, the 1-bit complex coding additions $\dot{0}_1 + \dot{1}_1 = \dot{1}_2$ and $\dot{1}_1 + \dot{0}_1 = \dot{3}_2$ produce different results. In this way, all digital states can be obtained with a uniform code distribution.

The new rules for the addition operations between two coding sets are given in Table 5.2. The performance of the new addition rules is validated in Figure 5.8d, in which the incident beam is equally deflected to four pencil beams without observable backscattering.

5.4.2 Performance Characterization

As known from the previous chapters, the 1-bit coding sequence 010101⋯ can only produce dual pencil beams that are symmetrical to the normal axis, which restricts its application scope in radar detection where two beams are required to radiate independently. In Figure 5.8e, we demonstrate a novel dual-beam

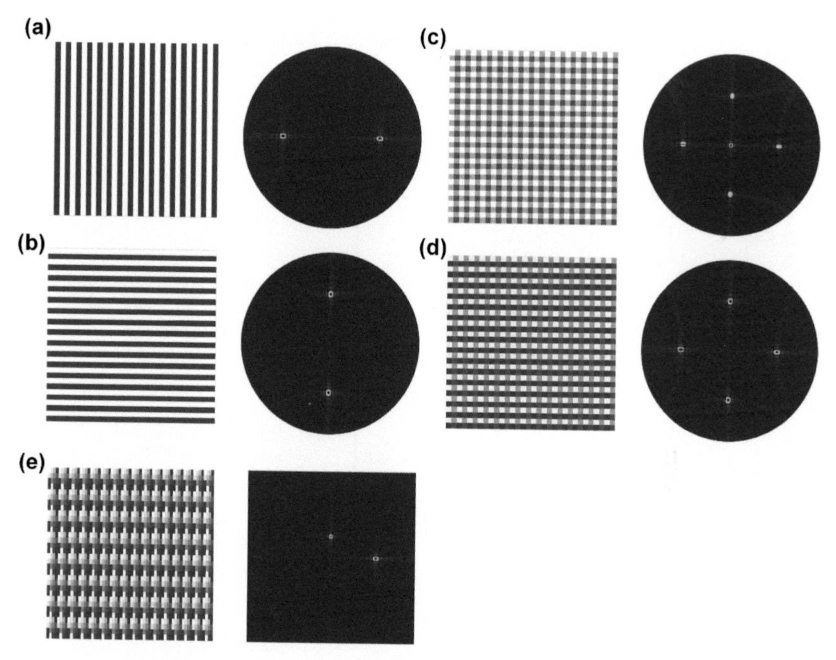

Figure 5.8 Coding schemes and their 3D and 2D scattering patterns calculated by FFT to illustrate the influence and solutions of indefinite coding addition . (a) Coding scheme of the coding sequence 00110011··· along the x direction. (b) Coding scheme of the coding sequence 00110011··· along the y direction. (c) Coding scheme by direct addition of the two initial coding sequences, in which the colors from light to dark represent the digital states from 0 to 3, showing the influence of indefinite coding addition. (d) New coding scheme by correcting the indefinite coding elements. (e) The addition operation of coding sequences 01230123··· along the x direction and 0011223300112233··· along the y direction.

radiator with 3-bit complex codes based on the addition theorem. Two pencil beams pointing at different directions in the *xoz* and *yoz* planes are combined into a quad-beam radiation pattern, with both radiation angle and intensity unperturbed. This example demonstrates the correctness of the new addition theorem in providing a fast and accurate coding strategy for creating such complicated radiation patterns.

Of course, the addition operations and convolution operations can work jointly to offer more versatile controls of the radiation patterns. For example, the convolution operation produces single-beam radiation at arbitrary directions, and the addition operation adds multiple single-beam radiations together to form an arbitrary multi-beam radiation pattern.

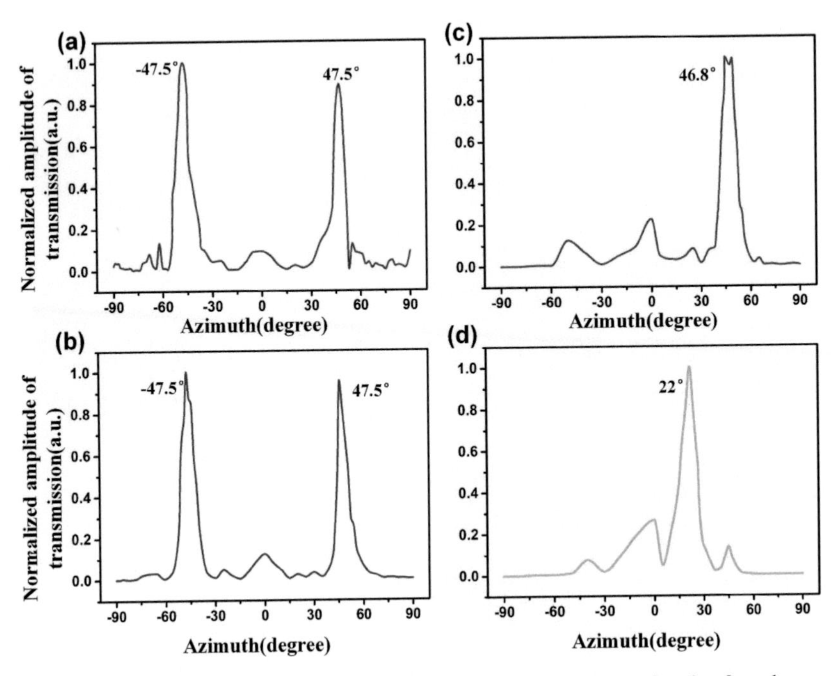

Figure 5.9 Experimental results. (a,b) 2D scattering patterns for the four-beam coding metasurface P_1 in the E-plane and H-plane, respectively. (c,d) 2D scattering patterns for the four-beam coding metasurface P_2 in the E-plane and H-plane, respectively.

5.4.3 Experiments and Measured Results

The performance of the addition theorem was experimentally demonstrated with a high-efficiency transmission-type coding metasurface, which consists of four square metallic patches separated by three dielectric substrates. A total of eight elements was obtained through numerical optimizations to form 3-bit complex digital codes, with the transmission coefficient of all elements exceeding 0.75. Two samples encoded with different coding patterns were fabricated to reflect the incident beam to two beams and four beams, respectively. It can be observed from Figures 5.9a and 5.9b that the reflected beams point at ±47.5° directions, which are consistent with the simulation results of ±49.18°. For the second case, the two dual-beam radiations point to directions of 46.8° and 22° (see Figures 5.9c and 5.9d), which again is in good agreement with the simulated values of 49.18° and 22.23°.

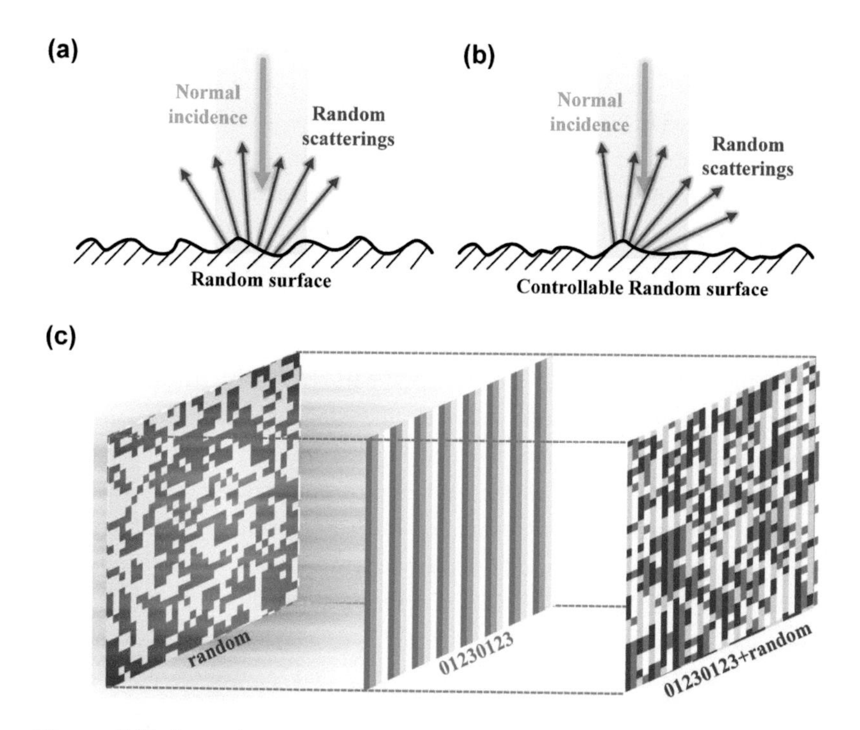

Figure 5.10 Controllable random surface based on coding metamaterial. (a) Conventional random surface. (b) Controllable random surface. (c) The procedure for generating the controllable random coding pattern (the right pattern) by mixing a random coding pattern (the left pattern) and a gradient coding pattern (the middle pattern).

5.5 Two Application Examples

5.5.1 Controllable Random Surface

We have demonstrated the phenomenon of random diffusions in both microwave and THz frequencies using the coding metasurfaces. It was recently discovered that GRS polynomials can be used to realize a well-diffused radiation pattern [159], in which the levels of random scattering are almost identical at all visible angles, leading to very low RCSs in all directions. However, neither the direction nor the intensity of the scattering waves of these random coding patterns can be predicted or controlled, as illustrated in Figure 5.10a, in which the random scattering is distributed around the normal axis of the random surface, resembling the well-known rough surfaces at the visible light spectra. As the coding metasurfaces can redirect the lights to arbitrary directions, a question naturally rises of whether we could control the direction of the random diffusion at will, for instance, to divert the random scatterings to the right-hand side of the normal axis, as illustrated in Figure 5.10b. The answer is yes.

Based on the convolution principle, we proposed the concept of a controllable random surface, which could manipulate the direction of random scattering and the level of diffusion by simply combining the periodic coding patterns with random coding patterns. Figure 5.10c shows the procedure for obtaining the controllable random coding pattern (the right pattern) from the mixture of a random coding pattern (the left pattern) and a gradient coding pattern (the middle pattern). As expected, the mixed coding pattern possesses the features of both the periodic and random coding patterns.

Two different examples are given in Figures 5.11a–d to demonstrate the exotic phenomenon of controllable random diffusions. Figure 5.11a shows the mixed coding pattern of the first example, which is obtained by adding a random coding pattern and a 2-bit periodic coding pattern "0 2 0 2 . . . " with super unit cell size of 3×3. Using the FFT method described in Section 5.1, we obtain the corresponding radiation pattern in Figure 5.11e, which displays two random diffusion regions beside the normal axis. Note that the central angles of the diffusion regions are exactly the anomalous reflection angles of the gradient

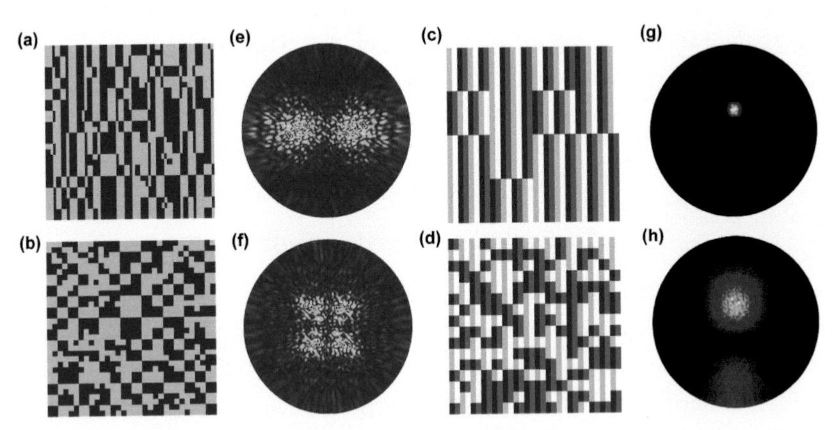

Figure 5.11 Simulated results of the controllable random surface. (a) The addition of a random coding pattern and a 2-bit periodic coding pattern "0 2 0 2 . . . " with super unit cell size of 3×3. (b) The addition of a random coding pattern and a 1-bit chessboard coding pattern. (c) The addition of a random coding pattern and a 2-bit periodic coding pattern "0 1 2 3 . . . " with super unit cell size of 16×16. (d) The addition of a random coding pattern and a 2-bit periodic coding pattern "0 1 2 3 . . . " with super unit cell size of 4×4. (e–h) The calculated radiation patterns corresponding to (a–d), respectively.

coding sequence, which are ($\theta = 30°$, $\varphi = 0°/180°$). As we add a chessboard coding pattern to the random pattern, we obtain the mixed coding pattern in Figure 5.11b, which generates four random diffusion regions around the angle of ($\theta = 32°$, $\varphi = \pm 45°/135°$), as illustrated in Figure 5.11f.

It is apparent from these examples that we could control the direction of random diffusions by selecting different periodic coding patterns. Now we further demonstrate that the level of diffusion – i.e. the region of random diffusion – can be arbitrarily tailored by changing the size of the super unit cell. The periodic coding pattern in Figures 42 c and d are the same with "0 1 2 3 0 1 2 3 . . . " gradient coding pattern, while the random coding patterns have super unit cell sizes of 16×16 and 4×4, respectively. Figures 5.11 g and h are the average results of radiation patterns calculated by 99 different random coding patterns. It can be found that the diffusion region in Figure 42 g is much smaller than that in Figure 42 h.

The smaller the size of the super unit cell, the more random coding digits will be included, and the more the uncertainty of the scattering will appear in a certain direction. Thus, the controllable random diffusion can be compared to, in some senses, the electron cloud in quantum mechanics, in which one never knows the exact position and velocity of the electron in the atom. Similarly, for the controllable random surface, it is unnecessary to control the direction of every random scattering, but necessary only to predict the probability of random scattering that appears at certain angles.

The concept of the controllable random surface can be extended to the microwave and optical frequencies, as well as to other physical domains (such as acoustics), leading to many revolutionary applications. For example, at microwave frequency, it can increase the imaging area of the single-sensor, single-frequency imaging technique, reviewed in Section 6.1. While at visible light spectrum, it promises a new type of diffusion material that could diffuse lights to the desired directions, which may be applied in warning boards, façade material of buildings, projection screens for cinema, etc. More interestingly, it can also be used to design novel acoustic diffusion materials with exotic diffusion properties.

5.5.2 Cone-Shaped Radiations with Arbitrary Opening Angles and Orientations

All previously mentioned digital coding patterns are designed to generate pencil beams pointing to arbitrary directions by adopting gradient coding patterns that vary along the x or y direction in the Cartesian coordinate. What happens if the gradient coding sequences vary along radial directions in the polar coordinate? To answer this question, we recently developed a coding metasurface that could generate ring-shaped radiation patterns with the EM energy confined within a small elevation angle and radiated equally in the azimuthal directions [158].

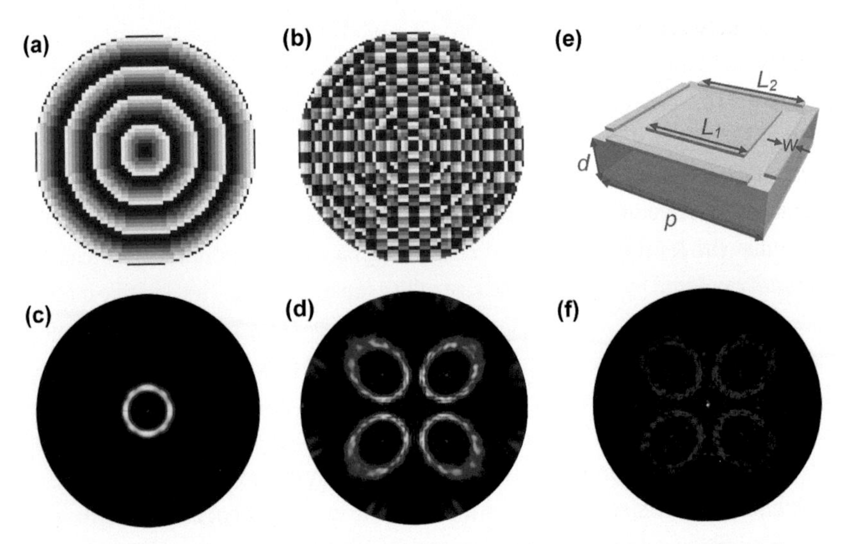

Figure 5.12 Cone-shaped radiation with arbitrary opening angle and orientation. (a) "0 1 2 3 4 5 6 7 0 1 2 3 4 5 6 7 . . . " (super unit cell size 2×2) varying along the radial direction. (b) The addition of the coding pattern in (a) with a chessboard coding pattern (super unit cell size 6×6). (c,d) The calculated radiation pattern corresponding to (a) and (b), respectively. (e) Structure of the unit cell. (f) Numerically simulated radiation pattern using the real structure.

Combining the principle of scattering pattern shift with the radially gradient coding patterns, we could manipulate the opening angle, direction and number of the ring-shaped radiation patterns.

To demonstrate the versatile ring-shaped radiation patterns using the principle of scattering pattern shift, two different coding patterns are given in Figures 2.10a and 2.10c, which are designed to generate a single-ring radiation pattern and a four-ring radiation pattern. The first coding pattern in Figure 5.12a is a radially gradient coding pattern – "0 1 2 3 4 5 6 7 0 1 2 3 4 5 6 7 . . . " (super unit cell size 2×2) – which generates a ring-shaped radiation pattern with its center pointing at the normal axis (see Figure 5.12b). If we add a chessboard coding pattern (super unit cell size 6×6) to the coding pattern in Figure 5.12a, we obtain the mixed coding pattern in Figure 5.12c, which exhibits four ring-shaped radiation patterns with their center pointing at (φ = ±45°/±135°, θ = 36.1°), as shown in Figure 5.12d.

To implement the proposed ring-shaped coding metamaterial, we designed a novel coding particle as presented in Figure 5.12e, which is superior to the square patch design in Ref. 140 in the following aspects. First, four metallic strips are added at the edge of each coding particle to reduce the undesired EM coupling between adjacent coding particles having different geometrical parameters. Second, the crystal quartz substrate features much lower loss ($\varepsilon_r = 4.4$; $\delta = 0.0004$) than the polyimide, thus supporting a high reflectance of more than 0.97 from 0.2 to 0.3 THz.

The thickness of the crystal quartz substrate is only 60 μm, which can be prepared by grinding and polishing a 500 μm-thick crystal quartz substrate. We built the coding pattern (see Figure 5.12b) with real structure in the CST Microwave Studio, and the simulated result (see Figure 5.12f) is in good agreement with the theoretical result (see Figure 5.12d), confirming the accurate response of the real structure. We remark that the novel coding metasurface with controllable ring-shaped radiation patterns may have potential applications in radar detection. As the ring-shaped radiation pattern radiates equally in the azimuthal direction, it has a faster scanning speed than the pencil-beam radar.

5.6 More Possible Signal Processing on Coding Metamaterials

The principle of scattering pattern shift is enabled by the digital characterization of coding metamaterials. It establishes a connection between the physical metamaterial and digital signal processing, allowing us to study and design metamaterials from the perspective of information science. We can expect many digital-signal-processing algorithms to be exploited and applied to the design of coding metamaterials, which could significantly reduce the complexity and simplify the process of metamaterial design.

For example, a fractional Fourier transform can be employed to calculate the near-field distribution of a coding metamaterial, in which the fractional order determines the distance of the calculation plane to the coding metamaterial. Some other algorithms like wavelet transform, low-pass filter, high-pass filter and even integration/differentiation operations can be possibly performed to the coding pattern to allow more flexible manipulations to the radiation patterns and to realize unprecedented functionalities.

6 New Imaging Systems Based on Programmable Metasurfaces

6.1 Single-Sensor and Single-Frequency Microwave Imaging Systems

Imaging has always been an important research topic, as well as a popular application in the field of electromagnetics. Over the past decades, many theories and technologies have been developed to increase the speed and quality of imaging. One of the interesting techniques employs only one sensor [189] to serve as both transmitter and receiver, thus significantly reducing the number of detectors in the array source in most imaging approaches. Based on this imaging technique, the sensor needs to generate different radiation patterns to collect sufficient data for wavefront reconstruction, or one needs to solve inverse scattering problems in the imaging plane.

Reconfigurable spatial light modulators (SLMs) were reported to realize a reflection-type single-pixel imaging in the THz regime [190]. Later, it was shown that 2D holographic metasurfaces can be used to reconstruct planar objects in the microwave frequency [191,192]. One of the key technologies behind the single-sensor imaging techniques is random modulators or masks [136,193,194], in which the transmission property of each constituent unit cell will be changed at different frequencies or different bias voltages. The programmable metamaterial with dynamically varying radiation patterns becomes a good candidate for implementing the sensor in the single-sensor and single-frequency imaging systems [195].

In 2016, we developed a single-sensor, single-frequency microwave imager based on the transmission-type 2-bit programmable metasurface. The schematic of the imaging system is presented in Figure 6.1a, and is mainly composed of a 2-bit transmission-type programmable metasurface and a target at the imaging plane. The programmable metasurface, connected to a vector-network analyzer (VNA), illuminates the target with dynamically changing random radiation patterns, and at the same time, receives the signal reflected back from the target in each measurement. To obtain an accurate image reconstruction of the object, the measurement needs to be repeated a sufficient number of times, normally not less than the number of pixels of the image. We take a very simple example with 5×5 = 25 meta-elements in the programmable metamaterial. Then a total number of 25 measurements was required to build up the matrix equation, each time with a different radiation pattern.

We briefly describe the principle of the imaging technique from a mathematical perspective. For a given radiation pattern, the measured signal can be defined as $V = \left(V_i^{(p)}\right)$, p=1, 2, ..., P, which is related to the original object-area vector $\sigma = (\sigma_i)$, i=1, 2, ..., N, through a generalized system response matrix $G = \left(G_{pj}\right)$, p=1,2,,,,,P, j=1, 2, ..., N. The relationship among V, G and σ is described by the following equation [195]:

$$\begin{bmatrix} V^{(1)} \\ V^{(2)} \\ \vdots \\ V^{(P)} \end{bmatrix} = \begin{bmatrix} G_{11} & G_{12} & \cdots & G_{1N} \\ G_{21} & G_{22} & & \\ & & \ddots & \\ G_{P1} & & & G_{PN} \end{bmatrix} \begin{bmatrix} \sigma_1 \\ \sigma_2 \\ \vdots \\ \sigma_N \end{bmatrix} \quad (6.1)$$

Here, as the image has 5×5 pixels, both P and N are equal to 5×5 = 25, resulting in a matrix, G, with 25×25 = 625 elements. To produce the image of the target, it is clear that we need to solve $\sigma = G^{-1}V$ with 25 data measurements (V).

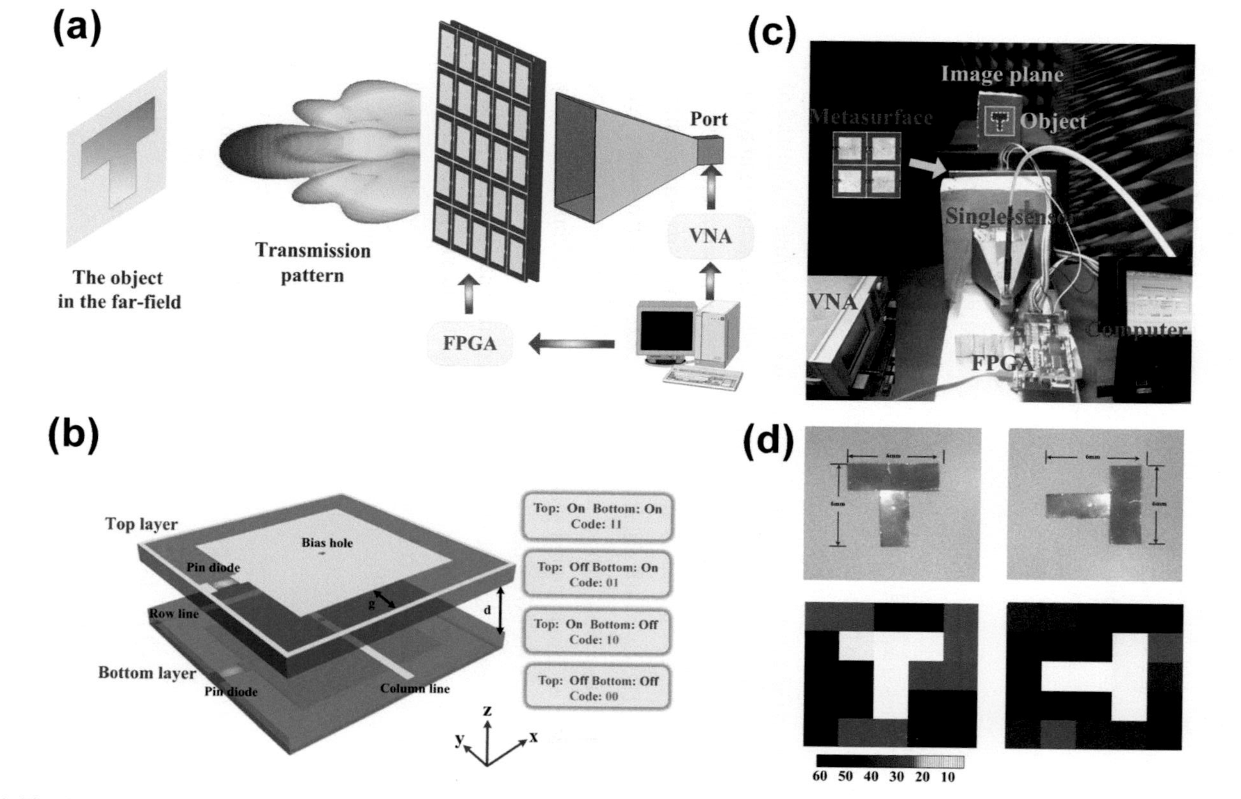

Figure 6.1 Single-sensor, single-frequency microwave imaging system. (a) Schematic illustration of the single-sensor, single-frequency microwave imaging system using programmable metamaterial. (b) Structure of the digital particle. (c) Experimental setup. (d) Experimentally reconstructed images of the T-shaped objects at 9.2 GHz.

Next, we make a brief introduction of the programmable metasurface that could generate the random radiation patterns in real time. Note that the programmable metasurface is different from the reflection-type programmable metamaterials presented in Section 4, which have some disadvantages in practical applications. For example, the radiation pattern suffers from non-negligible distortions due to the shielding effect of the feeding antenna placed in front of the programmable metasurface. Since the aperture of the horn antenna can be tightly attached to the transmission-type programmable metasurface, it takes less physical space than the reflection-type programmable metasurface.

Figure 6.1b illustrates the unit cell of the 2-bit transmission-type programmable metasurface working in the microwave frequency. To avoid the tremendous feeding lines required by the 2D programmable metasurface, we provided a smart design by controlling all digital particles by row and column [195]. To achieve the “row and column,” two pin-diodes (SMP 1320-079LF) were mounted on the top and bottom substrates and independently controlled by the column and row lines in the middle layer, which were connected to the central metal patches on the top and bottom layers through two metallic vias.

A metallic frame was added to the edge of the unit cell to serve as the common ground. The two F4B substrates with thickness of 1 mm were kept at a distance of 3 mm. By switching the two diodes to the ON and OFF states, we obtained four different digital states – “11”, “01”, “10” and “00” – as indicated in Figure 6.1b. Note that although the four different states cannot cover the 360° phase range, and the current feeding configuration does not allow independent control of each unit cell, it is sufficient for the single-sensor, single-frequency imaging system to generate dynamically random coding patterns in the considered frequency band of 9–10 GHz.

To experimentally demonstrate the performance of the single-sensor, single-frequency imaging system, we made a proof-of-principle prototype by fabricating a transmission-type programmable metamaterial with $10\times10 = 100$ elements. As a proof of the principle, the programmable metasurface was divided into $5\times5 = 25$ control units. Accordingly, the object was divided into $5\times5 = 25$ subareas, each having an area of 20×20 mm^2 (about $\lambda/2$ at 9 GHz). The experimental configuration of the imaging system is presented in Figure 6.1c, in which a computer automatically controls both VNA (Agilent N5230 C) and FPGA to generate a series of random radiation patterns at a frequency of 9.2 GHz and records the reflected signals.

One of the important issues in obtaining an excellent image is to determine the generalized system response matrix, which requires a careful calibration before the imaging experiment, and which can be accomplished by making a point-to-point scanning of a small metallic object placed at every subarea on the imaging plane.

A horizontal bar and a T-type object shown in Figure 6.1d were selected as the objects in the imaging tests. The lower panel of Figure 6.1d shows the experimentally reconstructed images of both objects at 9.2 GHz. The shape of the objects can be clearly identified from the imaging results, thereby verifying the effectiveness of the proposed transmission-type, programmable, metasurface-based imaging system.

6.2 Large-Aperture Imaging System with Compressed-Sensing Algorithm

In the previous section, the single-sensor, single-frequency imaging system is based on the full-matrix inversion, which limits the imaging scale with a small number of pixels. Here, we extend the method to realize a large-aperture single-sensor, single-frequency microwave imager based on a compressed sensing algorithm using the 1-bit programmable metasurface with 20×20 digital particles [196]. Owing to the column-row method, a total of 20+20 control lines is required, which significantly simplifies the design complexity of 20×20 lines required by the pixel-based method.

As such a control method does not allow independent control of every digital particle, the coding pattern cannot be arbitrarily selected. However, this does not affect the performance of the single-sensor, single-frequency microwave imager because the system does not require specific radiation patterns but simply the quasi-random radiation patterns.

We theoretically show that the proposed large-aperture computational single-shot imager can make a successful recovery of a sparse or compressible object by solving a sparsity-regularized convex optimization problem [196]. The following equations give a brief introduction to the kernel of the computational imaging algorithm. For a programmable metasurface illuminated by an x-polarized plane wave, the x-polarized radiation field in the far-field region can be expressed as

$$E^{(m)}(\mathbf{r}) = \sum_{n_x=1}^{N_x} \sum_{n_y=1}^{N_y} \widetilde{A}_{n_x,n_y}^{(m)} g(\mathbf{r}, \mathbf{r}_{n_x,n_y}) \tag{6.2}$$

in which $g(\mathbf{r}, \mathbf{r}_{n_x,n_y}) = \frac{e^{(jk_0|\mathbf{r}-\mathbf{r}_{n_x,n_y}|)}}{4\pi|\mathbf{r}-\mathbf{r}_{n_x,n_y}|}$ is the 3D Green's function in free space, m = 1,2, . . ., M indicates the number of coding patterns, and $\widetilde{A}_{n_x,n_y}^{(m)} = A_{n_x,n_y}^{(m)} e^{j\varphi_{n_x,n_y}^{(m)}}$ is the induced current on the (n_x,n_y)th unit cell of the programmable metasurface. The electric field at distance $\mathbf{r}$ can be obtained by performing the double summation in Equation (6.2) over all digital particles of the programmable metasurface. When the object placed at the imaging plane is illuminated by the wave field of Equation (6.2), the scattered electric field at distance r_d is written as

$$E^{(m)}(\mathbf{r}) = \sum_{n_x=1}^{N_x} \sum_{n_y=1}^{N_y} \widetilde{A}_{n_x,n_y}^{(m)} \widetilde{O}_{n_x,n_y} \tag{6.3}$$

$$\widetilde{O}_{n_x,n_y} = \int_V g(\mathbf{r}, \mathbf{r}_{n_x,n_y}) g(\mathbf{r}_d, \mathbf{r}) O(\mathbf{r}) d\mathbf{r} \tag{6.4}$$

where O(r) is the object function. Note that the attainable resolution of O(r) is on the order of $O(\lambda_R/D)$, in which λ is the operating wavelength, R is the observation distance, and D is the maximum size of the metasurface aperture. Equation (6.3) indicates that our target is to retrieve $N = N_x \times N_y$ unknowns $\{\widetilde{O}_{n_x,n_y}\}$ from M times of measurements. Typically, there is no unique solution for Equation (6.3) if $N > M$, due to the way it is intrinsically ill posed. This issue can be solved by searching for a sparsity-regularized solution to Equation (6.3), based on the assumption that the object $\widetilde{O}$ has a low-dimensional representation in certain transform domain denoted by Ψ. Now, the solution of Equation (6.3) can be achieved by solving the following sparsity-regularized optimization problem

$$min_{,O} \left[\frac{1}{2} \sum_{m=1}^{M} \left(E^{(m)} - \langle \ \widetilde{\mathbf{A}}^{(m)}, \widetilde{\mathbf{O}} \rangle \ + \gamma \left\| \Psi\left(\widetilde{\mathbf{O}} \right) \right\|_1 \right) \right] \tag{6.5}$$

in which γ is the balancing factor to trade off the data fidelity and sparsity prior.

Figure 6.2a shows the prototype of the large-aperture microwave imager, which includes 20×20 digital particles. Under the illumination of a horn antenna and using the external applied voltages, it could generate a series of random radiation patterns. The system is mainly composed of three parts: a horn antenna that launches the single-frequency microwave, a 1-bit programmable metasurface generating the sequentially compressed sensing random patterns, and a single sensor responsible for collecting the signals reflected from the targets, two metallic letters “P” and “K”. The reconstructed results of the single-sensor, single-frequency imaging system with 400 and 600 measurements are provided in the left and right panels in Figure 6.2b, respectively. The shapes of objects are clearly observed in the imaging results, demonstrating the feasibility of the proposed single-sensor imaging system, as well as the improvement of image quality by increasing the number of measurements.

6.3 Reprogrammable Coding Metasurface Hologram

Holography, as one of the most promising imaging techniques for recording the amplitude and phase information of lights to reconstruct the images of objects, has attracted tremendous attention owing to its wide applications in display, security, data storage, etc. Unfortunately, most conventional holographic

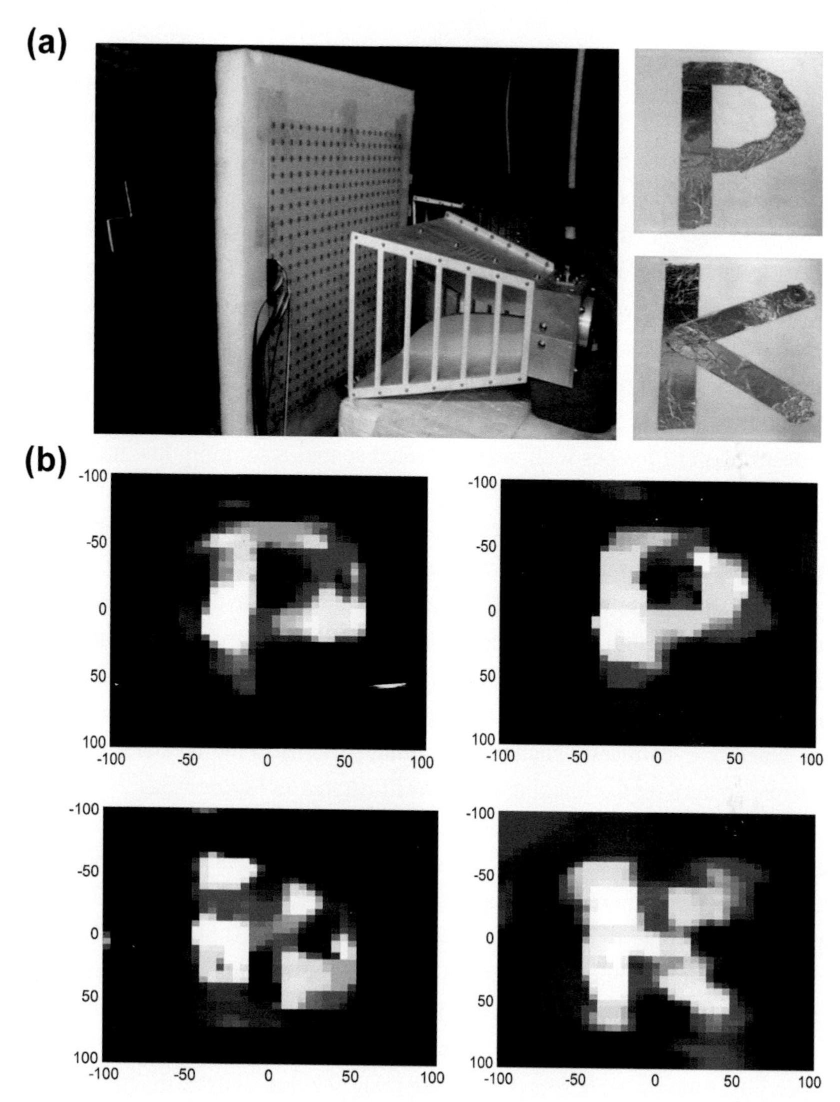

Figure 6.2 Large-aperture imaging system with compressed-sensing algorithm. (a) Photo of the fabricated large-aperture imaging system. (b) The reconstructed images of the two metallic letters "P" and "K" with 400 and 600 measurements.

systems suffer from limited resolution and poor image quality. In addition, the thickness of holograms is always comparable to the working wavelength, because the phase modulation relies on the spatial phase accumulation of lights passing through the bulky holograms.

Although a number of metasurface holograms have been reported over the past few years in the infrared and visible regimes [106,110,197], they can only generate static images because the phase and amplitude profiles of the passive metasurface are fixed once the metasurface is fabricated. With the discovery of active devices and materials such as thermal-sensitive phase change materials $Ge_2Sb_2Te_5$ [198,199], vanadium dioxide [200–202] and grapheme [203,204], active metasurfaces have recently become a research hotspot among scientists in fields from optics to material science.

Inspired by the dynamic feature of the programmable metasurface, we attempted to design a reprogrammable hologram with a 1-bit programmable metasurface [205], which features two main advantages compared with conventional holograms. First, owing to the reprogrammable deep subwavelength nature of the digital particles of the programmable metasurface, both phase and amplitude responses can be sampled with much higher resolution, thus enabling reconstruction of images with unprecedented spatial resolution, low noise and high precision. The second benefit afforded by the programmable metamaterial is the smaller unit cell, which could eliminate undesired diffractions and thus bring higher transmission or reflection efficiency.

The structure presented in Ref. 140 is slightly modified to design the metasurface hologram, as shown by the upper-left model in Figure 5.15a. By applying the corresponding voltages (indicated by the “0” and “1” digital states) to each digital particle through FPGA (or computer), the programmable metasurface can generate dynamically changing radiation patterns to the far field, forming different images on the imaging plane, as illustrated in Figure 6.3a.

One of the key issues in designing the metasurface hologram is how to obtain the correct coding patterns for the 1-bit programmable metasurface in order to generate the desired images on the imaging plane. By solving a combinatorial optimization problem, a modified Gerchberg-Saxton (GS) algorithm [206] is used to generate the binary phase profiles for the reprogrammable hologram, as opposed to the continuous-valued optimization problem encountered in the conventional GS algorithm. The current distributions of *J(1)* and *J(0)* – i.e. the coding patterns – are found by minimizing the non-convex function using the source inversion technique defined by the following equation [205],

$$Min\sum_{i=1}^{N_x N_y}\left(E^{des}(i)-\left|E(i)\right|\right)^2 \tag{6.6}$$

where E^{des} is the desired image. The optimization process starts with a random coding pattern and gradually converges to the desired coding pattern as the iteration process proceeds.

A sample was fabricated as a proof of concept for the microwave programmable metasurface hologram. A feeding antenna working from 6 to 14 GHz was employed to illuminate the metasurface with the *x*-polarized wave. The

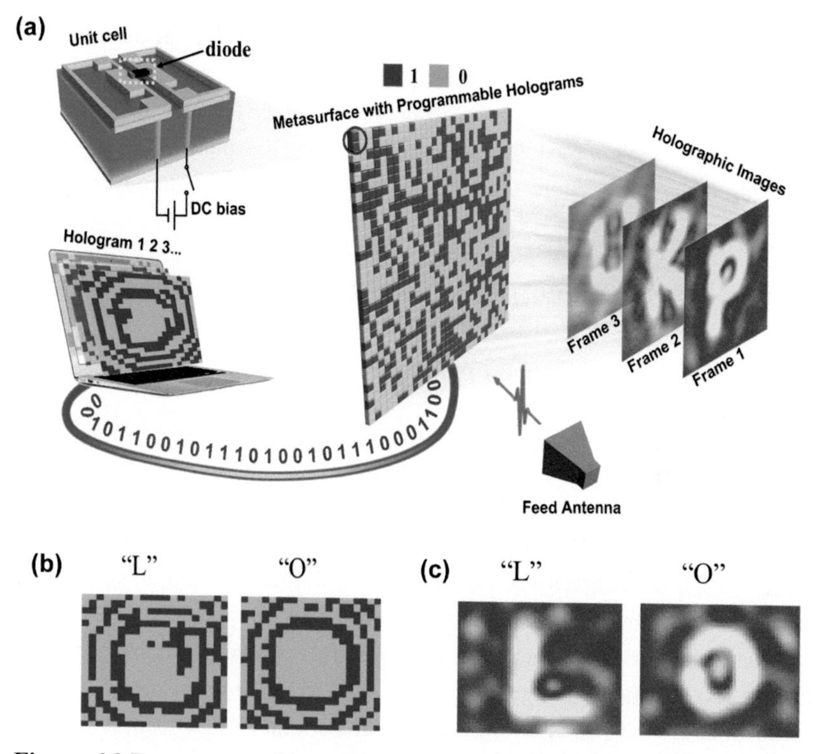

Figure 6.3 Reprogrammable coding metasurface hologram. (a) Schematic of the reprogrammable coding metasurface hologram. (b) Coding patterns for the letters "L" and "O". (c) Holographic images of the letters "L" and "O".

programmable metasurface, with a certain coding pattern at a certain time period, scattered the plane wave to the imaging plane to form the desired holographic image, which was then recorded by a standard waveguide probe with a resolution of 5×5 mm^2.

Sixteen English letters – "LOVE PKU! SEU! NUS!" – demonstrated the performance of the designed metasurface hologram. Figure 6.3b shows the coding patterns for the letters "L" and "O", in which the green and blue pixels represent the "0" and "1" states, respectively. The shapes of the letters "L" and "O" can be clearly identified from the holographic images measured at 400 mm away from the metasurface, as illustrated in Figure 6.3c. The overall efficiency, defined as the fraction of the incident energy that contributes to the holographic image, is estimated around 60%, and the signal-to-noise ratio (SNR), defined as the ratio of the peak intensity in the image to the standard deviation of the background noise, is around 10.

The metasurface hologram, as demonstrated, can project clear images when the observation plane is located 400–500 mm away from the metasurface, and could work in a 0.5 GHz bandwidth around a center frequency of 7.8 GHz. By making an adaptive adjustment to the coding patterns, clear holographic images can be obtained at further observation planes and other frequencies. Note that this is the first reprogrammable hologram designed in the microwave frequency, featuring dynamically projected images with exceptional resolution and high SNR.

7 Time-Domain Digital Coding and New-Architecture Wireless Communication System

7.1 Time-Domain Digital Coding Metasurface

In the digital coding and programmable metasurfaces mentioned in previous chapters, the digital coding is defined in the space domain. We have shown that different coding sequences in the space domain can be used to control the spatial beam behaviors of EM waves, such as steering a single beam, arbitrarily multiplying beam generations, wave diffusion, single and multiple cone-beam forming and even vortex beam manipulations. Here, we present a new kind of programmable metasurface to control the spectral (or harmonic) distribution of EM waves, in which the digital coding is defined in the time domain and the digital units are independently controlled with periodic time sequences through FPGA [207].

First, we analyzed the frequency spectra of the time-domain–coding programmable metasurface. For simplicity, let us assume a fixed-period time modulation. In this case, the reflected electric field can be expressed as a periodic function of time and defined over one period as a linear combination of scaled and shifted pulses. Under excitation with a monochromatic EM signal $E_i(f)$ with frequency f_c, the reflected signal from the time-domain digital coding metasurface can be written as

$$E_r(f) = a_0 E_i(f) + \sum_{k=1}^{\infty} \left[a_k E_i(f - kf_0) + a_{-k} E_i(f + kf_0) \right] \quad (7.1)$$

in which $f_0 = 1/T$, T is the period of the reflectivity function, and a_k is the complex Fourier series coefficient at kf_0:

$$a_k = \frac{1}{M} Sa\left(\frac{k\pi}{M}\right) \exp\left(-j\frac{k\pi}{M}\right) \cdot \sum_{m=0}^{M-1} \Gamma_m \exp\left(-j\frac{2km\pi}{M}\right) = UF \cdot TF \quad (7.2)$$

in which M is the length of the time-domain coding sequence in one period, Γ_m is the reflectivity at the interval $(m-1)\tau < t < m\tau$, $\tau = T/M$ is the pulse width, and

$$TF = \sum_{m=0}^{M-1} \Gamma_m \exp\left(-j\frac{2km\pi}{M}\right), UF = \frac{1}{M} Sa\left(\frac{k\pi}{M}\right) exp\left(-j\frac{k\pi}{M}\right) \quad (7.3)$$

Here, α_k can be written as the product of two terms: the time factor (TF) and the unit factor (UF), which respectively relate to the modulation signal Γ_m within different time slots and the Fourier series coefficient of the basic pulse with pulse width τ, repeated with the period T. From another perspective, UF and TF define the basic spectrum property of one pulse and the coding strategy, respectively. It is clear that a number of harmonics $f_c + kf_0$ can be generated by the time-domain programmable metasurface, with the amplitude and phase of each harmonic being controlled by the coefficients α_k based on an ad hoc modulation of TF.

7.2 Amplitude/Phase Modulation Time Coding Sequences

We compared two different modulations techniques, the amplitude modulation (AM) and phase modulation (PM), for the time-domain–coding-programmable metasurface in the manipulations of nonlinear components of the reflected waves. First, we calculated the spectral distribution of all harmonics under the monochromatic plane wave illumination. The spectra were found to be symmetrical with respect to the central frequency, with the $+k^{\text{th}}$ and $-k^{\text{th}}$ Fourier series components having equal amplitude. This is the inherent nature of the AM modulation, which is not desired in many communication systems because it causes energy waste and inefficient frequency utilization. In addition, it is not possible to fully suppress the zeroth-order harmonic (the central frequency) since the reflectivity should always be positive in the entire period.

To improve the modulation efficiency, we adopted phase modulation for the time-domain–coding-programmable metasurface, which gives asymmetric spectral responses. Here, the reflectivity was described as a signal with unity amplitude but a digital phase. Figure 7.1 shows the spectral intensity distributions of all harmonics under different time coding sequences. It is observed from Figures 7.1a and 7.1b that the zeroth-order harmonic is totally inhibited with a 1-bit coding sequence 010101 . . . ($M = 2$ and $T = 1$ μs), which is due to signal cancellation from anti-phase reflectivity in each period.

Figures 7.1c and 7.1d show the situation with 2-bit coding 00–01-00–01- . . . ., in which the reflectivity switches between 0° and 90° periodically. It can be found that the signals of even-order harmonics (except $k = 0$) are effectively

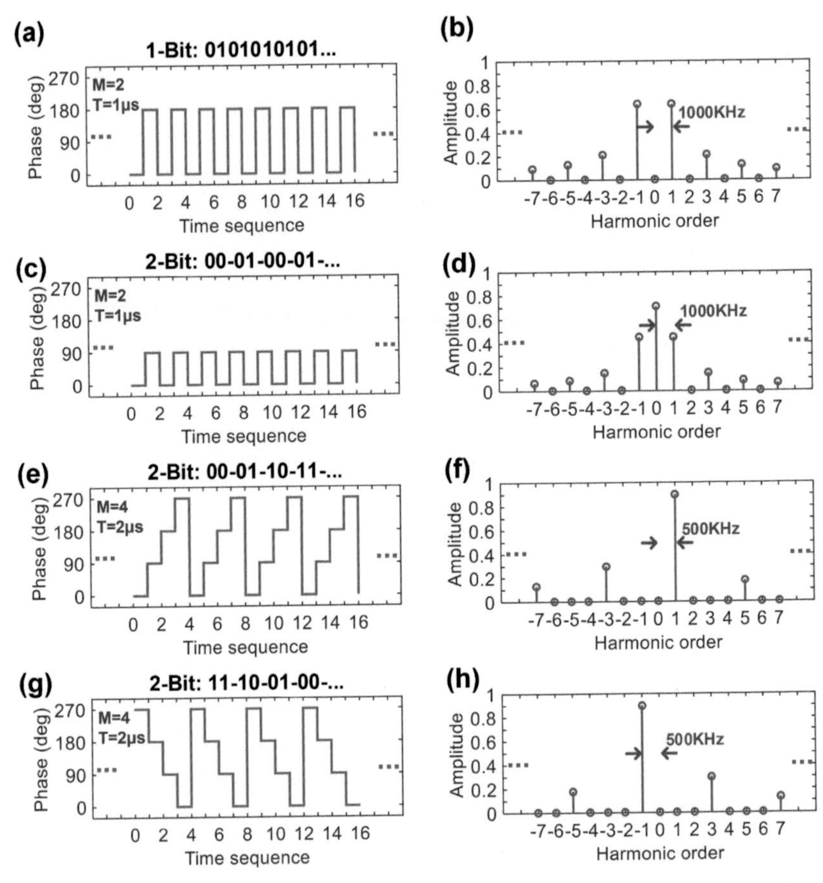

Figure 7.1 The calculated spectral intensities of the output harmonics under different PM modulations. (a,b) 1-bit PM coding 01010101 . . . with M = 2 and T = 1 μs. (c,d) 2-bit PM coding 00–01-00–01- . . . with M = 2 and T = 1 μs. **(e,f)** 2-bit PM coding 00–01-10–11- . . . with M = 4 and T = 2 μs. (g,h) 2-bit PM coding 11–10-01–00- . . . with M = 4 and T = 2 μs.

suppressed. Of course, we can further suppress other harmonics and create asymmetric energy distributions by considering higher-bit time-coding sequences with more phase states for the PM modulation. It is interesting to find that there is anti-symmetrical relation in the spectra between 2-bit coding sequences 00–01-10–11- . . . (Figures 7.1e and 7.1f) and 11–10-01–00- . . . (Figures 7.1g and 7.1h), which in mathematics is $a_{-k} = a_k^*$. Such an anti-symmetric phase ramp can lead to mirror transformations, which translates the original signals from the $+k^{\text{th}}$ harmonic to the corresponding $-k^{\text{th}}$ harmonic. Such a unique feature can be utilized to create the velocity illusion of a system moving in space, mimicking the Doppler shift [207].

7.3 Experiments on Nonlinear Harmonics Control

For experimental demonstration, we developed a PM time-domain–coding-programmable metasurface that consisted of 16×16 digital unit cells. Figure 7.2a shows the structure of the digital unit cell, which is composed of two rectangular patches connected by a varactor diode (SMV-2019) in substrate F4B with the dielectric constant $\varepsilon_r = 2.65(1-j0.001)$ and thickness 4 mm. A narrow slot (width = 0.15 mm) was etched on the back side of the substrate to bias the diodes. Another ultrathin vinyl electrical tape (3 M Temflex; thickness = 0.13 mm) backed by a metal layer was placed underneath the slots to eliminate the EM leakage through the slots [207]. The CST Microwave Studio was employed for the numerical optimization of the digital unit cell to work at the central frequency of 3.6 GHz. Voltages of 0 V and −9 V were applied to the varactor diodes to reach the "0" and "1" digital states.

Figure 7.2b illustrates the measured harmonic distributions modulated by the 1-bit time-domain digital coding 01010101 ... at different pulse durations. Consistent with the theoretical calculations, the incident energy at the central frequency shifted to the higher-order harmonics in measurements, with a significant drop (~15 dB) in the amplitude of the 0th-order harmonic. The $\pm 1^{st}$-order harmonics show nearly identical radiation patterns to those of the 0th-order harmonics. The harmonic energy level is insensitive to the pulse duration time τ, but is mostly determined by the time coding sequence.

For the 2-bit case, we applied bias voltages of 0, −6, −9 and −21 V to achieve the four digital states of "00", "01", "10" and "11", corresponding to phases 0°, 90°, 180° and 270°, respectively. The symmetry of the reflection spectra was broken because of the presence of time gradients, which can be confirmed from Figure 7.2c, in which the $\pm k^{th}$ harmonics exhibit obvious differences. The spectra can be mirrored by reversing the periodic coding sequence as 11–10-01–00, as shown in Figure 7.2d, as expected from the foregoing theoretical analysis. The radiation patterns measured in all three cases are in good agreement with the simulated results, as can be observed from Figures 7.2e and 7.2f.

The effective shifting of the carrier wave to the first harmonics enables the effect of an arbitrary Doppler shift, which can be used to deceive active radar systems by mimicking a moving target with echoed signals – namely, velocity illusion. For instance, a static time-domain digital coding metasurface with a frequency shift of ±156.25 KHz (see Figures 7.2c and 7.2d) may create an illusion of a moving object approaching (or receding) a radar operating at 3.6 GHz with a velocity of ±6.51 km/s. Dynamic Doppler shifts can be realized by

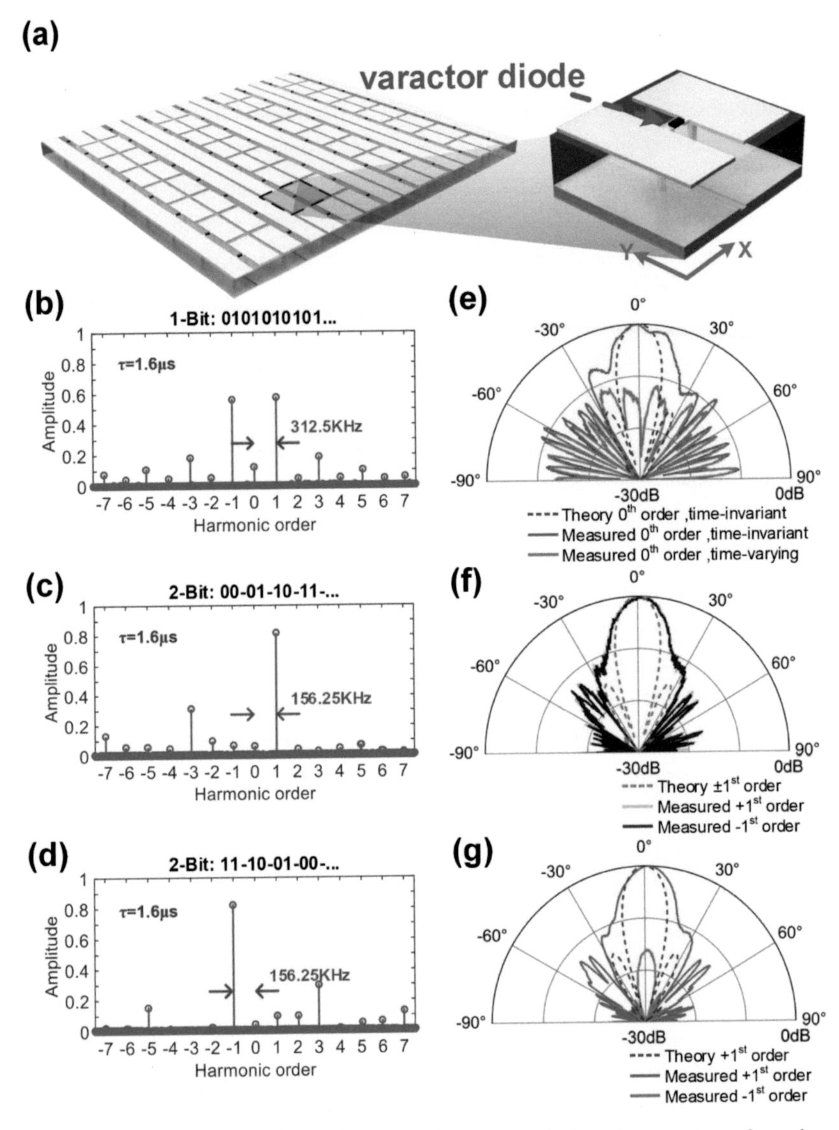

Figure 7.2 (a) Schematic of the time-domain digital coding metasurface, in which the inset shows the zoomed view of the meta-atom. (b) The measured spectral intensities of the harmonics under the 1-bit coding sequence 01010101 . . . at 3.6 GHz with the pulse duration $\tau = 1.6$ μs. (c,d) The measured spectral intensities of all harmonics under 2-bit coding sequences 00–01-10–11- . . . and 11–10-01–00- . . . at 3.6 GHz, respectively. (e) The measured H-plane scattering patterns of fundamental harmonic modulated by the 1-bit coding sequence 01010101 . . . (blue line) or not (red line). The theoretical radiation pattern of the 0th-order harmonic for the time-invariant

Caption for Figure 7.2 (cont.)

metasurface is also provided for comparison (dashed line). (f) The measured H-plane scattering patterns of the $+1^{st}$ (the green line) and -1st (the purple line) order harmonics modulated by the 1-bit coding sequence 01010101 . . . The dashed line shows the theoretical radiation pattern of $+1^{st}$-order harmonics for comparison. (g) The measured H-plane scattering patterns of the $+1^{st}$ (the blue line) and -1st (the red line) order harmonics under the 2-bit coding sequence 00–01-10–11- . . . The dashed line shows the theoretical radiation pattern of $+1^{st}$-order harmonics for comparison.

properly controlling the pulse width or coding sequence of the time-domain programmable metasurface, increasing the difficulty for radar systems to detect the movement of camouflaged targets.

7.4 New-Architecture Wireless Communication System

Modern wireless communication represents one of the most important technologies in the field of information engineering. The basic architecture of modern wireless communication systems is sketched in the upper diagram of Figure 7.3, which consists of a baseband module, DAC, up/down-convertor, carrier signal modulator, mixers, filters, low noise amplifiers, power amplifiers and antenna. This system architecture has not been challenged for decades. Here, we introduce a new binary frequency-shift keying (BFSK) communication system based on the time-domain–coding-programmable metasurface, as sketched in the lower diagram of Figure 7.3b, in which most of the modules in modern wireless communication systems can be simply replaced by the time-domain-coding digital metasurface and a feeding antenna [207]. Based on the capability of the time-domain programmable metasurface in manipulating the harmonics of a carrier wave, we developed a prototype of the new wireless communication system, in which the $\pm1^{st}$ harmonics components can be used for BFSK communication.

Figure 7.4a shows a schematic of the proposed BFSK system transmitter, which consists only of the time-domain programmable metasurface and a feeding antenna. The real-time BFSK signal transmission is carried out from the metasurface to soft-defined radio (SDR) receiver (NI USRP RIO 2943 R). The working flow can be divided into three steps. First, FPGA generates a bit stream (e.g. 01101001 . . .) to be transmitted (e.g. pictures and movies). Then, the bit streams are mapped to the corresponding coding sequences of the time-domain-coding metasurface, which are further extended

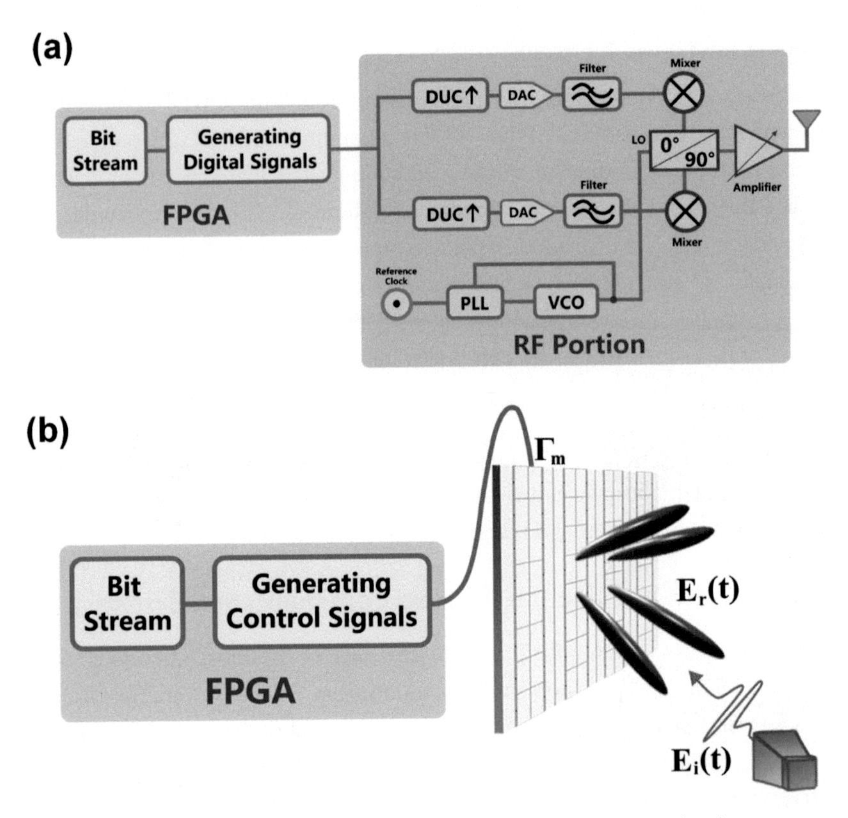

Figure 7.3 Architectures of (a) conventional wireless communication system and (b) new BFSK communication system (lower).

periodically to produce a pair of discrete frequencies in BFSK. Finally, the EM waves containing the digital information are radiated to free space [207].

The experiment was conducted in an anechoic chamber, as shown in Figure 7.4b. The BFSK wireless communication system transmitter was placed 6.25 m away from the receiver. All baseband algorithms for the BFSK receiver were performed on an SDR platform (NI USRP RIO 2943 R). The time-domain signal was transformed into the frequency domain through an FFT operation and then sent to the detecting diagram to determine the spectrum intensities. The upper path in the detecting diagram was responsible for the energy detection of frequency offset f_1 (related to bit "1"), while the lower path was responsible for the energy detection of frequency offset f_2 (related to bit "0"). When the power level detected in the upper path exceeds that of the lower path, the receiver will judge the current bit to be transmitted as "1", and vice versa.

The received bit stream will be recovered, grouped and sent for post-processing to obtain the final data. Table 7.1 shows the key parameters of the

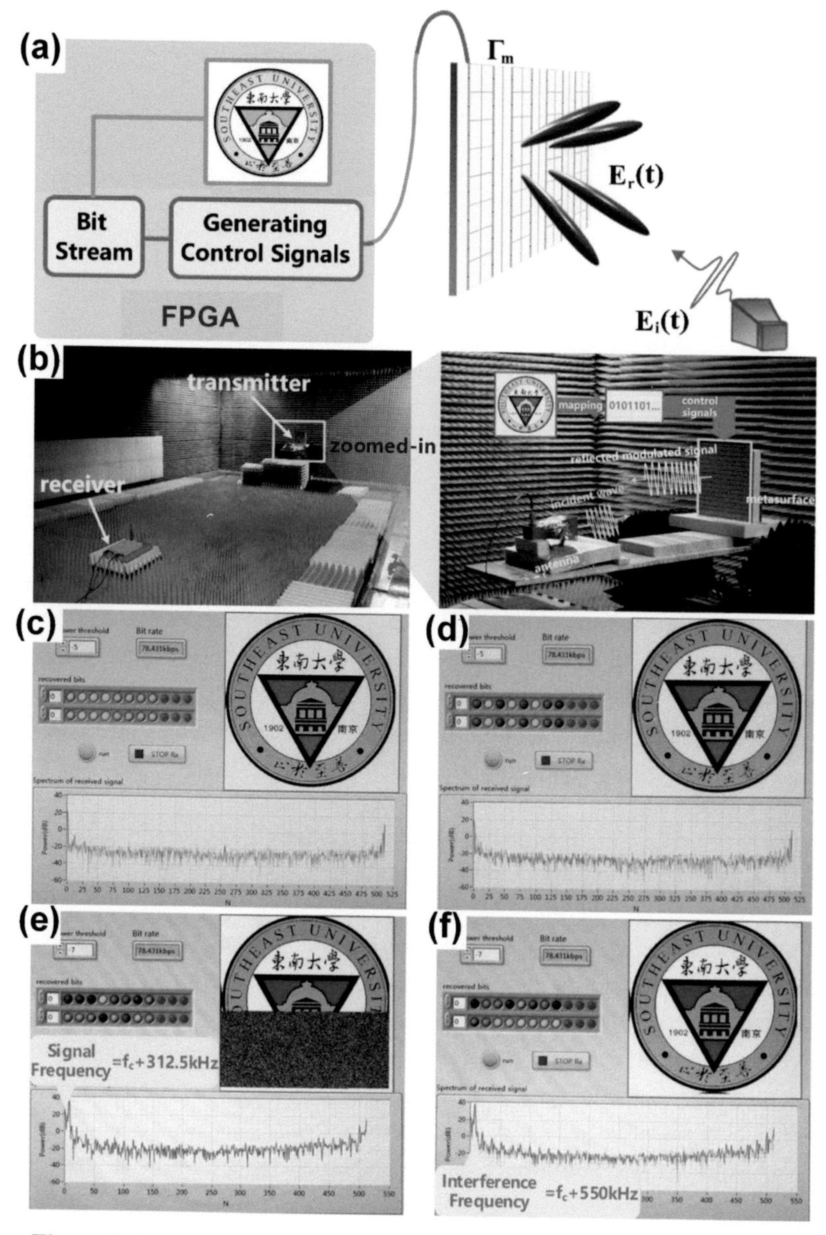

Figure 7.4 (a) Schematic of the proposed BFSK wireless communication system based on time-domain digital coding metasurface. (b) Experimental scenario of the BFSK wireless communication system with transmission process described on the right. (c,d) The received messages by the BFSK wireless communication system for different receiving angles $\alpha = 0^{\circ}$ and 30°, respectively. (e,f) Receiving process of the BFSK wireless communication system with an interference frequency at f_c+550 kHz, while the signal frequency is $f_c \pm 312.5$ kHz, to show robust anti-interference ability.

Table 7.1 The primary parameters of the BFSK wireless communication system based on the time-domain digital coding metasurface

Parameters	Values
Carrier frequency	3.6 GHz
Sampling rate	40 MS/s
FFT size	512
Frequency offset 1 (bit "1")	+312.5 KHz
Frequency offset 2 (bit "0")	−312.5 KHz
Message symbol duration	12.8 μs
Bit rate of transmission	78.125 Kbps

BFSK wireless communication system based on the time-domain digital coding metasurface. Figures 7.4c and 7.4d show the received images with receiving angles $\alpha = 0°$ and 30°, which clearly demonstrate the correctness of the system. We also evaluated the anti-interference performance of the new communication system with an interference frequency of f_c+550 kHz. The clearly received images shown in Figures 7.4e and 7.4f demonstrate the robustness of the system against the interference frequency close to the signal frequency.

8 Summary and Outlook

8.1 Summary of This Element

Information metamaterials have been rapidly developed and expanded to other areas since they were first proposed in 2014, breaking the boundaries of 3D metamaterials and 2D metasurfaces outlined by physics. In this short Element, we have reviewed many intriguing properties and applications of information metamaterials. Some theories and design strategies may have profound influence in both scientific and engineering communities.

For example, the anisotropic digital coding metamaterial is of extreme practical importance for designing ultrathin and flexible dual-image holograms, and possibly doubling the data storage density of the currently widely used optical disks. The structural flexibility of the reflection-type coding metamaterials could effectively reduce the radar cross-sections of objects, for example, the nose of an aircraft or radome for ground antennas, by covering them around the object surface. One of the pioneering achievements is the establishment of convolution theorem on the digital coding metasurface, which breaks the restriction of the scanning angle resolution supported by the purely gradient metasurfaces. This principle is very general and, if we are able to fabricate a sufficiently large digital coding metamaterial, we could realize a single-beam radiation pointing in arbitrary directions. The phenomenon of continuous beam scanning is also of fundamental interest and has promising applications in radar detection and imaging systems, particularly those requiring high scanning resolutions at low cost.

The digital coding representation of metamaterials not only presents new design strategies for complicated metamaterials but also results in a new class of programmable metamaterials with integration to FPGA, which stores all possible coding sequences to reach plenty of functionalities. Hence, the programmable metamaterials can be used to control EM waves, including the spatial beams and spectral distributions, in real-time programmable ways. More importantly, the digital coding and programmable metamaterials have built a bridge between the physical world and the digital world, which makes the metamaterials not only effective materials but also effective information-processing systems. At the end of this Element, we briefly introduced two important examples of system-level applications: the single-sensor, single-frequency microwave imaging system and the new-architecture wireless communication system. We remark that all concepts and design principles proposed in this Element can be applied, in principle, to any wavelengths of interest, even to acoustic waves.

8.2 Future Directions and Outlook

Information metamaterials establish a solid connection between the physical world and the digital world from which many existing algorithms in digital signal processing can be borrowed to realize new functionalities. Intriguingly, the digital coding metamaterials can be employed to process signals from the physical level. In 2014, metamaterials were reported to perform mathematical operations on the impinging wave, such as spatial differentiation, integration and convolution [208]. Compared with conventional lens-based optical signal and data processors, such a wave-based computing system is greatly miniaturized and potentially can be integrated with photonic and optical devices. However, the current system comprising of two GRIN lenses and a metasurface is still relatively bulky and can possess only a fixed computation function. By using the digital coding and programmable metamaterials, we expect that various mathematical operations to the incident field will be performed in a single metamaterial system.

For the programmable metamaterials at higher frequencies, the biggest problem is experimental implementation. In fact, enormous efforts have been devoted to the realization of active metamaterials in the THz and infrared frequencies over the past decade, for example, the LC-based tunable metamaterial [209], MEMS-based metamaterials for dynamic manipulation of infrared radiation [210] and grapheme-based metasurface for broadband THz modulation [211]. Other approaches are controlled by laser beam [199,212–214] or temperature [202,215]. However, most of them are focused on the improvement of transmission/reflection amplitudes, and few on the independent control of metamaterial elements. Representative work on the experimental realization of independent control include the THz spatial light modulator, which consists of multiple unit cells being independently controlled by an applied voltage [136], and the gate-tunable metasurface for dynamic electrical control of the phase and amplitude of the infrared beam [204].

One of the recent developments that may lead to the programmable metamaterial is associated with the use of microfluidics structures, which takes precise control and manipulation of fluids that are geometrically constrained to a small, typically sub-millimeter, scale. By changing the filling factor of each mercury-filled split-ring resonator equipped with pneumatic valves, one can manipulate the phase of the transmitted microwave [216]. With a deliberately designed microfluidic network, each resonator could be individually controlled, thus enabling dynamic wavefront shaping of the transmission phase. A series of similar works was reported recently, including a wide-angle multifunctional polarization converter with arm length of each L-shaped Galinstan resonator dynamically controlled using the microfluidic channels [217], and an adaptable metasurface comprising a periodic array of liquid-metal ring-shaped resonators for dynamic anomalous reflections [218].

Another possible method to control the state of digital particles is to utilize mechanical movement, as the structural modification can result in an effective change in both phase and amplitude of the incident wave. Such mechanically controlled programmable metamaterials have obvious advantages. First, they can easily provide the 2π phase range with high reflection/refraction amplitude, as compared to the electrically controlled programmable metamaterials, which are typically very lossy due to the intrinsic loss of RF switches. Another benefit is that the states of all digital particles can be maintained even when the power is switched off, which is not the case for the electrically controlled programmable metamaterials, because the digital states of unit cells can only be kept under the existence of external bias voltage. In Ref. 219, a circularly polarized reconfigurable reflect array antenna comprising 15×15 elements was developed for steering microwave (X band) to the desired directions with high aperture efficiency. The concentric dual split rings in each unit cell are mechanically rotated by a micro-motor to obtain the continuous 360° phase coverage with negligible loss, and the entire array is experimentally demonstrated to have a wide scan range of ±60°.

Although the reconfigurable reflect array antennas have been made to manipulate the microwave [219–222], fabricating the mechanically controlled programmable metamaterial in the infrared and visible light ranges is still very challenging because current micro-fabrication techniques can hardly scale down the mechanical structure by several orders of magnitude to the optical range. Recently, an electromechanically reconfigurable plasmonic metamaterial has been suggested, operating in the near-infrared by tuning the gap between two parallel strings using electrostatic forces [223]. Owing to the elastic properties of a nanoscale-thickness dielectric membrane and the nanoscale electrostatic forces, the device can provide continuous modulations of optical signals with a megahertz bandwidth. Unfortunately, this design can only manipulate the transmission amplitude with a limited modulation step of ~5%, but cannot support independent controls of individual elements. Further study is expected to realize THz and optical programmable metasurfaces.

In previous chapters, we demonstrated that space-coding digital metamaterials have powerful capabilities in controlling the special beams of EM waves, while time-coding digital metamaterials can be used to manipulate the spectral distributions of harmonics, accompanying the process of digital messages in real time in the physical level. In order to increase the controlling ability of EM waves and information capacity simultaneously, space-time-coding digital information metamaterials can manipulate both spatial beams and spectral distributions of EM waves and process high-dimensional digital messages [224,225]. The space-time-coding digital metamaterials have shown some exciting features, including harmonic beam steering, extremely low RCS reductions by wave diffusions in both spatial and frequency domains and even breaking the symmetry in space-time domain [226]. More new physical phenomena and information behaviors remain to be explored in the space-time-coding digital metamaterials, which is a good direction for future work.

Currently, most of the digital coding metamaterials (in the space and time domains) have focused on the phase modulations. Several works have touched the amplitude digital coding metamaterials, but they are still limited to the passive situations [227–230]. In the future, amplitude-phase joint coding digital metamaterials will receive much attention because they can incorporate more digital signal-processing algorithms to enhance information capacity. Besides, the integration with amplifier devices in the digital particles will bring more flexibilities and capabilities in controlling EM waves, engineering nonlinearity, amplifying reflections or transmissions and editing the digital information [231–234].

Here, we share with the readers in Figure 8.1 the technology road map we envision for the future development of information metamaterials. Now, we

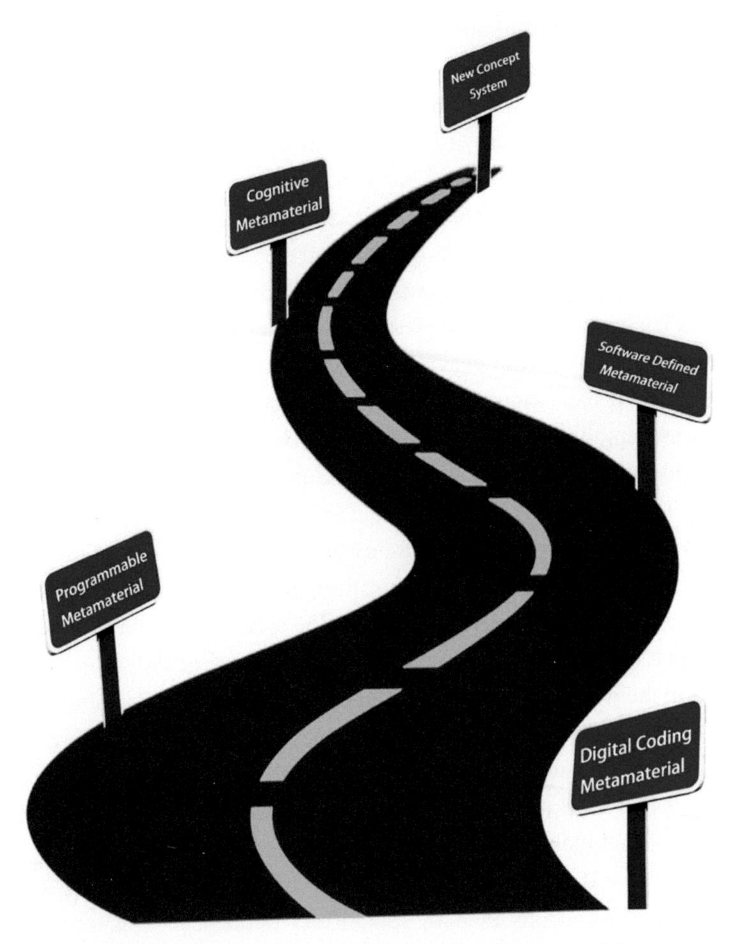

Figure 8.1 Roadmap for information metamaterials

have passed by the first and second states on the road map, digital coding metamaterial and programmable metamaterial, which have progressed in the past several years. However, many other possible functionalities may exist if we further exploit the potential combination between digital coding metamaterials and information science. The third milestone on the roadmap is the software-defined metamaterials, in which the coding pattern of the programmable metamaterials for realizing specific functions can be instantly obtained from fast algorithms, software, or even machine-learning methods integrated in FPGA [235–237]. Such a real-time response is of particular importance in applications, and is also the basic requirement for information metamaterials to evolve into the fourth stage, cognitive metamaterial. The biggest difference between software-defined metamaterial and cognitive metamaterial is that the latter

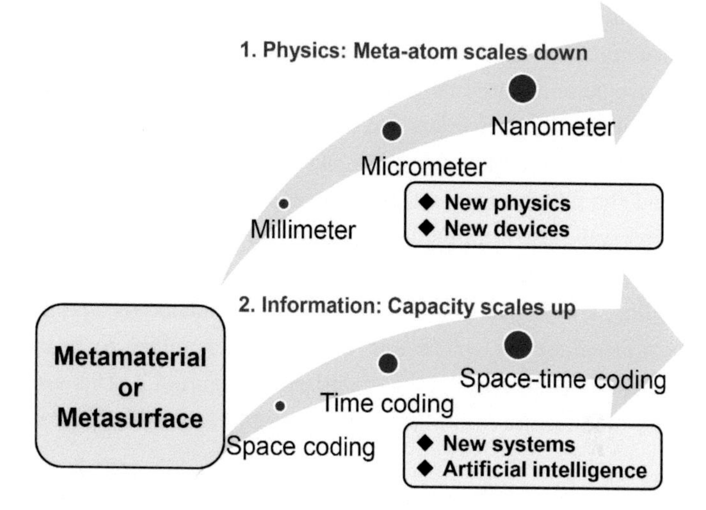

Figure 8.2 Developing trend of metamaterials during the past 20 years

should be able to sense the outer EM environment to enable a more exact response in real time [238]. Besides, the cognitive metamaterial should be supported by pre-stored databases and machine-learning algorithms in FPGA to gain the ability of self-learning. In this way, cognitive metamaterial can resemble a chameleon. We believe that future information metamaterial is expected to have intelligent features like self-sensing, self-learning, self-adaptation and self-decision.

Although many new physics and exotic phenomena will be found as we follow the technology road map, the major task and the ultimate aim of the information metamaterials are to realize new information systems with engineering applications. For example, a new type of wireless digital communication systems has recently been built based on the programmable metamaterial [239] by delivering information to the far-field region via dynamically changing radiation patterns. From a more simplistic perspective, it can be stated that the programmable metamaterial in the new wireless communication system serves as a complete communication model that integrates most of the digital models and microwave components (such as the D/A converters, mixers, filters, amplifiers, and antennas) used in the conventional wireless communication system, thus directly sending the digital data to the receivers in a fully digital manner.

In retrospect of the development of metamaterials during the past 20 years, there is a trend to make meta-atoms smaller and smaller, from the millimeter scale to micrometer and nanometer scales, as shown in the upper portion of

Figure 8.2, leading to microwave metamaterial, optical metamaterial and quantum metamaterial. As part of this developing trend, more new physical findings will be discovered and more interesting devices will be developed (e.g. antennas, lenses, polarization converters and beam splitters).

In addition to the trend toward micro-nano scopes with more new physics, another trend should advance toward macroscope to reach system-level applications. Apparently, the information metamaterial is a good candidate in this direction. With space-domain coding, time-domain coding and space-time-domain coding, the digital and programmable metamaterials involve more and more information capacities and have stronger ability to process the information, as illustrated in the lower portion of Figure 8.2. This will result in many new-architecture information systems and even artificial intelligence systems.

References

1. T. Brunet, A. Merlin, B. Mascaro, K. Zimny, J. Leng, O. Poncelet, C. Aristegui, O. M. Monval, Soft 3D acoustic metamaterial with negative index, *Nat. Mater.* 14, 384–388 (2015).
2. L. Zigoneanu, B.-I. Popa, S. A. Cummer, Three-dimensional broadband omnidirectional acoustic ground cloak, *Nat. Mater.* 13, 352–355 (2014).
3. B. I. Popa, S. A. Cummer, Non-reciprocal and highly nonlinear active acoustic metamaterials, *Nat. Commun.* 5, 3398 (2014).
4. S. A. Cummer, J. Christensen, A. Alù, Nature-controlling sound with acoustic metamaterials, *Nature Reviews Materials* 1, 16001 (2016).
5. T. J. Cui, D. R. Smith, R. Liu, *Metamaterials*, Springer, 2010.
6. N. Engheta, R. W. Ziolkowski, *Metamaterials: Physics and engineering explorations*, Wiley, 2006.
7. D. R. Smith, D. C. Vier, T. Koschny, C. M. Soukoule, Electromagnetic parameter retrieval from inhomogeneous metamaterials, *Phys. Rev. E* 71, 036617 (2005).
8. R. W. Ziolkowski, E. Heyman, Wave propagation in media having negative permittivity and permeability, *Phys. Rev. E* 64, 056625 (2001).
9. D. Schurig, J. J. Mock, D. R. Smith, Electric-field-coupled resonators for negative permittivity metamaterials, *Appl. Phys. Lett.* 88, 041109 (2006).
10. J. B. Pendry, A. J. Holden, D. Robbins, W. J. Stewart, Magnetism from conductors and enhanced nonlinear phenomena, *IEEE Trans on Microw Theory & Tech* 47(11), 2075–2084 (1999).
11. A. N. Grigorenko, A. K. Geim, H. F. Gleeson, Y. Zhang, A. A. Firsov, I. Y. Khrushchev, J. Petrovic, Nanofabricated media with negative permeability at visible frequencies, *Nature* 438, 335–338 (2005).
12. D. R. Smith, W. J. Padilla, D. C. Vier, S. C. Nemat-Nasser, S. Schultz, Composite medium with simultaneously negative permeability and permittivity, *Phys. Rev. Lett.* 84, 4184 (2000).
13. R. A. Shelby, D. R. Smith, S. Schultz, Experimental verification of a negative index of refraction, *Science* 292(5514), 77–79 (2001).
14. J. B. Pendry, Negative refraction makes a perfect lens, *Phys. Rev. Lett.* 85 (18), 3966 (2000).
15. C. M. Soukoulis, M. Wegener, Past achievements and future challenges in the development of three-dimensional photonic metamaterials, *Nat. Photon.* 5, 523–530 (2011).

16. N. I. Zheludev, The road ahead for metamaterials, *Science* 328, 582–583 (2010).
17. J. B. Pendry, D. Schurig, D. R. Smith, Controlling electromagnetic fields, *Science* 312, 1780–1782 (2006).
18. U. Leonhardt, Optical conformal mapping, *Science* 312 1777–1780 (2006).
19. W. X. Jiang, C. W. Qiu, T. C. Han, Q. Cheng, H. F. Ma, S. Zhang, T. J. Cui, Broadband all-dielectric magnifying lens for far-field high-resolution imaging, *Adv. Mater.* 25(48), 6963–6968 (2013).
20. B. D. F. Casse, W. T. Lu, Y. J. Huang, E. Gultepe, L. Menon, S. Sridhar, Super-resolution imaging using a three-dimensional metamaterials nanolens, *Appl. Phys. Lett.* 96, 023114 (2010).
21. R. Liu, C. Ji, J. Mock, J. Chin, T. Cui, D. Smith, Broadband ground-plane cloak, *Science* 323, 366–369 (2009).
22. H. F. Ma, T. J. Cui, Three-dimensional broadband ground-plane cloak made of metamaterials, *Nat. Commun.* 1(21) (2010).
23. W. X. Jiang, H. F. Ma, Q. Cheng, T. J. Cui, Illusion media: Generating virtual objects using realizable metamaterials, *Appl. Phys. Lett.* 96, 121910 (2010).
24. Y. Lai, J. Ng, H. Chen, D. Han, J. Xiao, Z. Zhang, C. T. Chan, Illusion optics: The optical transformation of an object into another object, *Phys. Rev. Lett.* 102(25), 253902 (2009).
25. W. X. Jiang, C. W. Qiu, T. Han, S. Zhang, T. J. Cui, Creation of ghost illusions using wave dynamics in metamaterials, *Adv. Func. Mater.* 23 (32),4028–4034 (2013).
26. J. B. Pendry, A chiral route to negative refraction, *Science* 306(5700), 1353–1355 (2004).
27. J. Yao, Z. Liu, Y. Liu, Y. Wang, C. Sun, G. Bartal, A. M. Stacy, X. Zhang, Optical negative refraction in bulk metamaterials of nanowires, *Science* 321(5891), 930 (2008).
28. E. Cubukcu, K. Aydin, E. Ozbay, S. Foteinopoulou, C. M. Soukoulis, Negative refraction by photonic crystals, *Nature* 423, 604–605 (2003).
29. C. Zhang, T. J. Cui, Negative reflections of electromagnetic waves in a strong chiral medium, *Applied Physics Letters* 91, 194101 (2007).
30. X. M. Yang, X. Y. Zhou, Q. Cheng, H. F. Ma, T. J. Cui, Diffuse reflections by randomly gradient index metamaterials, *Optical Letters* 35(6),808–810.
31. S. H. Lee, C. M. Park, Y. M. Seo, C. K. Kim, Reversed Doppler effect in double negative metamaterials, *Phys. Rev. Lett.* 81, 241102 (2010).
32. Q. Cheng, T. J. Cui, W. X. Jiang, B. G. Cai, Reversed Doppler effect in double negative metamaterials, *New. J. Phys.* 12, 063006 (2010).

33. E. E. Narimanov, A. V. Kildishev, Optical black hole: Broadband omnidirectional light absorber, *Appl. Phys. Lett.* 95, 041106 (2009).
34. H. F. Ma, T. J. Cui, Three-dimensional broadband and broad-angle transformation-optics lens, *Nat. Commun.* 1, 124 (2010).
35. R. W. Ziolkowski, A. Erentok, Metamaterial-based efficient electrically small antennas, *IEEE Trans. Antenna & Propagat.* 54(7), 2113–2130.
36. H. F. Ma, X. Chen, H. S. Chen, X. M. Yang, W. X. Jiang, T. J. Cui, Experiments on high-performance beam-scanning antennas made of gradient-index metamaterials, Appl. *Phys. Lett.* 95, 094107 (2009).
37. X. Chen, H. F. Ma, X. Y. Zhou, W. X. Jiang, T. J. Cui, Three-dimensional broadband and high-directivity lens antenna made of metamaterials, *J. Appl. Phys.* 110, 044904 (2011).
38. M. Q. Qi, W. X. Tang, T. J. Cui, A broadband Bessel beam launcher using metamaterial lens, *Sci. Rep.* 5, 11732 (2015).
39. Q. Cheng, H. F. Ma, T. J. Cui, Broadband planar Luneburg lens based on complementary metamaterials, *Appl. Phys. Lett.* 95, 101901 (2009).
40. B. Zhou, Y. Yang, T. J. Cui, Beam-steering Vivaldi antenna based on partial Luneburg lens constructed with composite materials, *J. Appl. Phys.* 110, 084908 (2011).
41. Z. L. Mei, J. Bai, T. M. Niu, T. J. Cui, A half Maxwell fish-eye lens antenna based on gradient-index metamaterials, *IEEE Trans. Antenna & Propagat.* 60(1), 398–401 (2012).
42. H. F. Ma, B. G. Cai, T. X. Zhang, Y. Yang, W. X. Jiang, T. J. Cui, Three-dimensional gradient-index materials and their applications in microwave lens antennas, *IEEE Trans. Antenna & Propagat.* 61(5), 2561–2569 (2013).
43. X. Q. Lin, T. J. Cui, J. Y. Chin, X. M. Yang, Q. Cheng, R. Liu, Controlling electromagnetic waves using tunable gradient dielectric metamaterial lens, *Appl. Phys. Lett.* 92, 131904 (2018).
44. N. I. Landy, S. Sajuyigbe, J. J. Mock, D. R. Smith, W. J. Padilla, Perfect metamaterial absorber, *Phys. Rev. Lett.* 100, 207402 (2008).
45. S. Liu, H. B. Chen, T. J. Cui, A broadband terahertz absorber using multi-layer stacked bars, *Appl. Phys. Lett.* 106, 151601 (2015).
46. H. Tao, C. M. Bingham, A. C. Strikwerda, D. Pilon, D. Shrekenhamer, N. I. Landy, K. Fan, X. Zhang, W. J. Padilla, R. D. Averitt, Highly flexible wide angle of incidence terahertz metamaterial absorber: Design, fabrication, and characterization, *Phys. Rev. B* 78, 244103 (2008).
47. H. Li, L. H. Yuan, X. P. Shen, Q. Cheng, T. J. Cui, Ultrathin multiband gigahertz metamaterial absorbers, *J. Appl. Phys.* 110, 014909 (2011).

48. X. P. Shen, T. J. Cui, J. Zhao, H. F. Ma, W. X. Jiang, H. Li, Polarization-independent wide-angle triple-band metamaterial absorber, *Opt. Express* 19(10),9401–9407 (2011).
49. N. K. Grady, J. E. Heyes, D. R. Chowdhury, Y. Zeng, M. T. Reiten, A. K. Azad, A. J. Taylor, D. A. R. Dalvit, H. T. Chen, Terahertz metamaterials for linear polarization conversion and anomalous refraction, *Science* 340 (6138),1304–1307 (2013).
50. J. Y. Chin, M. Lu, T. J. Cui, Metamaterial polarizers by electric-field-coupled resonators, *Appl. Phys. Lett.* 93, 251903 (2008).
51. Y. Ye, S. He, 90° polarization rotator using a bilayered chiral metamaterial with giant optical activity, *Appl. Phys. Lett.* 96, 203501 (2010).
52. X. M. Yang, X. Y. Zhou, Q. Cheng, H. F. Ma, T. J. Cui, Diffuse reflections by randomly gradient index metamaterials, *Optics Letters* 35(6), 808–810 (2010).
53. A. V. Kildishev, A. Boltasseva, V. M. Shalaev, Planar photonics with metasurfaces, *Science* 339 (6125), 1232009 (2013).
54. N. Yu, P. Genevet, M. A. Kats, F. Aieta, J.-P. Tetienne, F. Capasso, Z. Gaburro, Light propagation with phase discontinuities: Generalized laws of reflection and refraction, *Science* 334, 333–337 (2011).
55. E. H. Khoo, E. P. Li, K. B. Crozier, Plasmonic wave plate based on subwavelength nanoslits, *Optics Letters* 36, 2498–2500 (2011).
56. Y. Zhao, A. Alù, Manipulating light polarization with ultrathin plasmonic metasurfaces, *Phys. Rev. B* 84, 205428 (2011).
57. A. Pors, M. G. Nielsen, R. L. Eriksen, S. I. Bozhevolnyi, Broadband focusing flat mirrors based on plasmonic gradient metasurfaces, *Nano Lett.* 13, 829–834 (2013).
58. P. Genevet, N. Yu, F. Aieta, J. Lin, M. A. Kats, R. Blanchard, M. O. Scully, Z. Gaburro, F. Capasso, Ultra-thin plasmonic optical vortex plate based on phase discontinuities, *Appl. Phys. Lett.* 100, 013101 (2012).
59. Y. Liu and X. Zhang, Metasurfaces for manipulating surface plasmons, *Appl. Phys. Lett.* 103, 141101 (2013).
60. F. Aieta, P. Genevet, M. A. Kats, N. Yu, R. Blanchard, Z. Gaburro, F. Capasso, Aberration-free ultrathin flat lenses and axicons at telecom wavelengths based on plasmonic metasurfaces, *Nano Lett.* 12, 4932–4936 (2012).
61. B. A. Munk, *Frequency selective surfaces: Theory and design*, Wiley, 2000.
62. J. Huang, J. A. Encinar, Introduction to *Reflectarray antenna*, Wiley, 2007.
63. F. Capolino, Part I: Super-resolution, Applications of metamaterials, in P. A. Belov, ed., *Metamaterials handbook*, Taylor & Francis, 2009 2.8–2.10.

64. F. Capolino, Theory and phenomena of metamaterials, in *Metamaterials handbook*, Taylor & Francis, 2009.
65. H. Tao, N. I. Landy, C. M. Bingham, X. Zhang, R. D. Averitt, W. J. Padilla, A metamaterial absorber for the terahertz regime: Design, fabrication and characterization, *Opt. Express* 16, 7181–7188 (2008).
66. G. Dayal, S. A. Ramakrishna, Broadband infrared metamaterial absorber with visible transparency using ITO as ground plane, *Opt. Express* 22, 15104 (2014).
67. X. L. Liu, T. Starr, A. F. Starr, W. J. Padilla, Infrared spatial and frequency selective metamaterial with near-unity absorbance, *Phys. Rev. Lett.* 104, 207403 (2010).
68. M. B. Pu, C. G. Hu, M. Wang, C. Huang, Z. Y. Zhao, C. T. Wang, Q. Feng, X. G. Luo, Design principles for infrared wide-angle perfect absorber based on plasmonic structure, *Opt. Express* 19, 17413 (2011).
69. K. Aydin, V. E. Ferry, R. M. Briggs, H. A. Atwater, Design principles for infrared wide-angle perfect absorber based on plasmonic structure, *Nat. Commun*. 2, 517 (2011).
70. W. G. Yeo, N. K. Nahar, K. Sertel, Far-IR multiband dual-polarization perfect absorber for wide incident angles, *Microwave Opt. Technol. Lett.* 55, 632 (2013).
71. Z. Y. Fang, Y. R. Zhen, L. R. Fan, X. Zhu, P. Nordlander, Tunable wide-angle plasmonic perfect absorber at visible frequencies, *Phys. Rev. B* 85, 245401 (2012).
72. Y. Ma, Q. Chen, J. Grant, S. C. Saha, A. Khalid, D. R. S. Cumming, A terahertz polarization insensitive dual band metamaterial absorber, *Opt. Lett*. 36, 945–947 (2011).
73. B. X. Wang, X. Zhai, G. Z. Wang, W. Q. Huang, L. L Wang, A novel dual-band terahertz metamaterial absorber for a sensor application, *J. Appl. Phys.* 117, 014504 (2015).
74. S. Liu, J. C. Zhuge, S. J. Ma, H. B. Chen, D. Bao, Q. He, L. Zhou, T. J. Cui, A bi-layered quad-band metamaterial absorber at terahertz frequencies, *J. Appl. Phys.* 118, 245304 (2015).
75. R. Yahiaoui, J. P. Guillet, F. D. Miollis, P. Mounaix, Ultra-flexible multi-band terahertz metamaterial absorber for conformal geometry applications, *Optics Letters* 38, 4988–4990 (2013).
76. X. P. Shen, Y. Yang, Y. Zang, J. Q. Gu, J. G. Han, W. L. Zhang, T. J. Cui, Triple-band terahertz metamaterial absorber: Design, experiment, and physical interpretation, *Appl. Phys. Lett.* 101, 154102 (2012).

77. J. F. Zhu, Z. F. Ma, W. J. Sun, F. Ding, Q. He, L. Zhou, Y. G. Ma, Ultra-broadband terahertz metamaterial absorber, *Appl. Phys. Lett.* 105, 021102 (2014).
78. J. Grant, Y. Ma, S. Saha, A. Khalid, D. R. S. Cumming, Polarization insensitive, broadband terahertz metamaterial absorber, *Optics Letters* 36, 3476–3478 (2011).
79. N. F. Yu, F. Aieta, P. Genevet, M. A. Kats, Z. Gaburro, F. Capasso, A Broadband, Background-free quarter-wave plate based on plasmonic metasurfaces, *Nano Lett.* 12, 6328–6333 (2012).
80. N. Yu, P. Genevet, F. Aieta, M. Kats, R. Blanchard, G. Aoust, J. P. Tetienne, Z. Gaburro, F. Capasso, Flat optics: Controlling wavefronts with optical antenna metasurfaces, *IEEE J. Select. Topics Quantum Electron.* 19, 4700423 (2013).
81. R. Blanchard, G. Aoust, P. Genevet, N. Yu, M. A. Kats, Z. Gaburro, F. Capasso, Modeling nanoscale V-shaped antennas for the design of optical phased arrays, *Phys. Rev. B* 85, 155457 (2012).
82. M. Kats, P. Genevet, G. Aoust, N. Yu, R. Blanchard, F. Aieta, Z. Gaburro, F. Capasso, Giant birefringence in optical antenna arrays with widely tailorable optical anisotropy, *Proc. Natl Acad. Sci. USA* 109, 12364–12368 (2012).
83. F. Aieta, A. Kabiri, P. Genevet, N. Yu, M. A. Kats, Z. Gaburro, F. Capasso, Reflection and refraction of light from metasurfaces with phase discontinuities, *J. Nanophoton.* 6, 063532 (2012).
84. Y. M. Yang, W. Y. Wang, P. Moitra, I. I. Kravchenko, D. P. Briggs, J. Valentine, Dielectric meta-reflectarray for broadband linear polarization conversion and optical vortex generation, *Nano Lett.* 14, 1394–1399 (2014).
85. Z. W. Xie, X. K. Wang, J. S, Ye, S. F. Feng, W. F. Sun, T. Akalin, Y. Zhang, Spatial terahertz modulator, *Sci. Rep.* 3, 3347 (2013).
86. W. T. Chen, K. Y. Yang, C. M. Wang, Y. W. Huang, G. Sun, I. Chiang, C. Y. Liao, W. L. Hsu, H. T. Lin, S. L. Sun, L. Zhou, A. Q. Liu, D. P. Tsai, High-efficiency broadband meta-hologram with polarization-controlled dual images, *Nano Lett.* 14, 225–230 (2014).
87. X. H. Ling, H. Liu, J. H. Teng, A. Danner, S. Zhang, C. W. Qiu, Visible-frequency metasurface for structuring and spatially multiplexing optical vortices, *Adv. Mater.* 28, 2533–2539 (2016).
88. L. L. Huang, X. Z. Chen, H. Mühlenbernd, H. Zhang, S. M. Chen. B. F. Bai, Q. F. Tan, G. F. Jin, K. W. Cheah, C. W. Qiu, J. S. Li, T. Zentgraf, S. Zhang, Three-dimensional optical holography using a plasmonic metasurface, *Nat. Commun.* 4, 2808 (2013).

89. Y. Montelongo, J. O. T. Tenorio, C. Williams, S. Zhang, W. I. Milne, T. D. Wilkinson, Plasmonic nanoparticle scattering for color holograms, *Proc. Natl. Acad. Sci.* 111, 12679–12683 (2014).
90. W. M. Ye, F. Zeuner, X. Li, B. Reineke, S. He, C. W. Qiu, J. Liu, Y. T. Wang, S. Zhang, T. Zentgraf, Spin and wavelength multiplexed nonlinear metasurface holography, *Nat. Commun.* 7, 11930 (2016).
91. Z. J. Ma, S. M. Hanham, P. Allbella, B. Ng, H. T. Lu, Y. D. Gong, S. A. Maier, M. H. Hong, Terahertz all-dielectric magnetic mirror metasurfaces, *ACS Photon.* 3, 1010–1018 (2016).
92. M. Khorasaninejad, W. T. Chen, R. C. Devlin, J. Oh, A. Y. Zhu, F. Capasso, Metalenses at visible wavelengths: Diffraction-limited focusing and subwavelength resolution imaging, *Science* 352, 1190–1194 (2016).
93. M. L. Tseng, H.-H., Hsiao, C. H. Chu, M. K. Chen, G. Sun, A-Q Liu, D. P. Tsai, Metalenses: advances and applications, *Adv. Opt. Mater.* 6(18), 1800554 (2019).
94. S. Wang, P. C. Wu, V.-C. Su, T.-C. Lai, M.-K. Chen, H. Y. Kuo, B. H. Chen, Y. H. Chen, T.-T. Huang, J.-H. Wang, R.-M. Lin, C.-H. Kuan, T. Li, Z. Wang, S. Zhu, D. P. Tsai, A broadband achromatic metalens in the visible, *Nat. Nanotech.* 13(3), 227 (2018).
95. W. Zang, Q. Yuan, R. Chen, L. Li, T. Li, X. Zou, G. Zhang, Z. Chen, S. Wang, Z. Wang, S. N. Zhu, Chromatic dispersion manipulation based on metalenses, *Adv. Mater.* 1904935 (2019).
96. Z. Bomzon, V. Kleiner, E. Hasman, Formation of radially and azimuthally polarized light using space-variant subwavelength metal stripe gratings, *Appl. Phys. Lett.* 79, 1587–1589 (2001).
97. Z. Bomzon, V. Kleiner, E. Hasman, Pancharatnam-Berry phase in space-variant polarization-state manipulations with subwavelength gratings, *Optics Letters* 26, 1424–1426 (2001).
98. F. Monticone, N. M. Estakhri, A. Alù, Full control of nanoscale optical transmission with a composite metascreen, *Phys. Rev. Lett.* 110, 203903 (2013).
99. C. Pfeiffer, A. Grbic, Metamaterial Huygens' surfaces: Tailoring wave fronts with reflectionless sheets, *Phys. Rev. Lett.* 110, 197401 (2013).
100. S. A. Schelkunoff, Some equivalence theorems of electromagnetics and their application to radiation problems, *Bell Syst. Tech. J.* 15, 92–112 (1936).
101. A. Alù, Mantle cloak: Invisibility induced by a surface, *Phys. Rev. B* 80, 245115 (2009).
102. P.-Y. Chen, A. Alù, Mantle cloaking using thin patterned metasurfaces, *Phys. Rev. B* 84, 205110 (2011).

103. P.-Y. Chen, C. Argyropoulos, A. Alù, Broadening the cloaking bandwidth with non-Foster metasurfaces, *Phys. Rev. Lett.* 111, 233001 (2013).
104. S. Liu, H. X. Xu, H. C. Zhang, T. J. Cui, Tunable ultrathin mantle cloak via varactor-diode-loaded metasurface. *Opt. Express* 22, 13403–13417 (2014).
105. S. Liu, H. C. Zhang, H. X. Xu, T. J. Cui, Nonideal ultrathin mantle cloak for electrically large conducting cylinders. *Journal of the Optical Society of America A* 31, 2075–2082 (2014).
106. S. Larouche, Y.-J., Tsai, T., Tyler, N. M., Jokerst, D. R., Smith, Infrared metamaterial phase holograms. *Nat. Mater.* 11, 450–454 (2012).
107. X. J. Ni, A. V. Kildishev, V. M. Shalaev, Metasurface holograms for visible light, *Nat. Commun*. 4, 2807 (2013).
108. G. Zheng, H. Muhlenbernd, M. Kenney, G. Li, T. Zentgraf, S. Zhang, Metasurface holograms reaching 80% efficiency, *Nat. Nanotechnology*, 10, 308–312 (2015).
109. Y. W. Huang, W. T. Chen, W. Y. Tsai, P. C. Wu, C. M. Wang, G. Sun, D. P. Tsai, High-efficiency broadband meta-hologram with polarization-controlled dual images, aluminum plasmonic multicolor meta-hologram, *Nano Lett*. 15, 3122–3127 (2015).
110. L. Huang, H. Muhlenbernd, X. Li, X. Song, B. Bai, Y. Wang, T. Zentgraf, Broadband hybrid holographic multiplexing with geometric metasurfaces, *Adv. Mat.* 27, 6444–6449 (2015).
111. H. Wong, K. W. Cheah, E. Y. B. Pun, S. Zhang, X. Z. Chen, Helicity multiplexed broadband metasurface holograms, *Nat. Commun*. 6, 8241 (2015).
112. X. Fang, H. Ren, M. Gu, Orbital angular momentum holography for high-security encryption, *Nat. Photon.* 14(12), 102–108 (2020).
113. A. Minovich, D. N. Neshev, D. A. Powell, I. V. Shadrivov, Y. S. Kivshar, Tunable fishnet metamaterials infiltrated by liquid crystals, *Appl. Phys. Lett.* 96, 193103 (2010).
114. M. Decker, C. Kremers, A. Minovich, I. Staude, A. E. Miroshnichenko, D. Chigrin, D. N. Neshev, C. Jagadish, Y. S. Kivshar, Electro-optical switching by liquid-crystal controlled metasurfaces, *Opt. Express* 21, 8879–8885 (2013).
115. M. A. Kats, R. Blanchard, P. Genevet, Z. Yang, M. M. Qazilbash, D. N. Basov, S. Ramanathan, F. Capasso, Thermal tuning of mid-infrared plasmonic antenna arrays using a phase change material, *Optics Letters* 38, 368–370 (2013).
116. G. Biener, A. Niv, V. Kleiner, E. Hasman, Geometrical phase image encryption obtained with space-variant subwavelength gratings, *Optics Letters* 30, 1096–1098 (2005).

117. Y. Yirmiyahu, A. Niv, G. Biener, V. Kleiner, E. Hasman, Vectorial vortex mode transformation for a hollow waveguide using Pancharatnam-Berry phase optical elements, *Optics Letters* 31, 3252–3254 (2006).
118. W. M. Zhu, A. Q. Liu, T. Bourouina, et.al. Microelectromechanical Maltese-cross metamaterial with tunable terahertz anisotropy. *Nat. Commun.* 3, 1274 (2012).
119. J. Y. Ou, E. Plum, L. Jiang, N. I. Zheludev. Reconfigurable photonic metamaterials, *Nano. Lett.* 11, 2142 (2011).
120. M. Lapine, I. V. Shadrivov, D. A. Powell, Y. S. Kivshar. Magnetoelastic metamaterials, *Nat. Mater.* 11, 30 (2012).
121. J. Zhang, K. F. Macdonald, N. I. Zheludev. Nonlinear dielectric optomechanical metamaterials. *Light: Sci. Appl.* 2, e96 (2013).
122. A. Kuzyk, R. Schreiber, H. Zhang, et al., Reconfigurable 3D plasmonic metamolecules. *Nat. Mater.* 13, 826–866 (2014).
123. G. Biener, A. Niv, V. Kleiner, E. Hasman, Space-variant polarization scrambling for image encryption obtained with subwavelength gratings, *Opt. Commun.* 261, 5–12 (2006).
124. T. S. Kasirga, Y. N. Ertas, M. Bayindir, Microfluidics for reconfigurable electromagnetic metamaterials. *Appl. Phys. Lett.* 95, 214102 (2009).
125. M. W. Klein, C. Enkrich, M. Wegener, S. Linden, Second-harmonic generation from magnetic metamaterials, *Science* 313, 502–504 (2006).
126. V. K. Valev, A. V. Silhanek, N. Verellen, W. Gillijns, P. van Dorpe, O. A. Aktsipetrov, G. A. E. Vandenbosch, V. V. Moshchalkov, T. Verbiest, Asymmetric optical second-harmonic generation from chiral-shaped gold nanostructures, *Phys. Rev. Lett.* 104, 127401 (2010).
127. H. Husu, B. K. Canfield, J. Laukkanen, B. Bai, M. Kuittinen, J. Turunen, M. Kauranen, Chiral coupling in gold nanodimers, *Appl. Phys. Lett.* 93, 183115 (2008).
128. P.-Y. Chen, C. Argyropoulos, A. Alù, Enhanced nonlinearities using plasmonic nanoantennas, *Nanophotonics* 1, 221–233 (2012).
129. M. Lapine, I. V. Shadrivov, Y. S. Kivshar, Nonlinear metamaterials, *Rev. Mod. Phys.* 86, 1093–1123 (2014).
130. Y. Yao, M. A. Kats, P. Genevet, N. F. Yu, Y. Song, J. Kong, F. Capasso, Broad electrical tuning of graphene-loaded plasmonic antennas. *Nano Lett.* 13, 1257–1264 (2013).
131. Z. Feng, Y. Wang, A. E. Schlather, Z. Liu, P. M. Ajayan, F. J. G. de Abajo, P. Norlander, X. Zhu, N. J. Halas, Active tunable absorption enhancement with graphene nanodisk arrays, *Nano Lett.* 14, 299–304 (2014).

132. H. Yan, X. Li, B. Chandra, G. Tulevski, Y. Wu, M. Freitag, W. Zhu, P. Avouris, F. Xia, Tunable infrared plasmonic devices using graphene/insulator stacks, *Nature Nanotechnol.* 7, 330–334 (2012).
133. S. Thongrattanasiri, F. H. L. Koppens, F. J. G. de Abajo, Complete optical absorption in periodically patterned graphene, *Phys. Rev. Lett.* 108, 047401 (2012).
134. J. A. H. van Nieuwstadt, M. Sandtke, R. H. Harmsen, F. B. Segerink, J. C. Prangsma, S. Enoch, L. Kuipers, Strong modification of the nonlinear optical response of metallic subwavelength hole arrays, *Phys. Rev. Lett.* 97, 146102 (2006).
135. H.-T. Chen, W. J. Padilla, M. J. Cich, A. K. Azad, R. D. Averitt, A. J. Taylor, A metamaterial solid-state terahertz phase modulator, *Nat. Photon.* 3, 148–141 (2009).
136. W. L. Chan, H. T. Chen, A. J. Taylor, I. Brener, M. J. Cich, D. M. Mittleman, A spatial light modulator for terahertz beams, *Appl. Phys. Lett.* 94, 213511 (2009).
137. D. R. Smith, S. Schultz, P. Markoš, C. M. Soukoulis, Determination of effective permittivity and permeability of metamaterials from reflection and transmission coefficients, *Phys. Rev. B* 65, 195104 (2002).
138. R. P. Liu, T. J. Cui, D. Huang, B. Zhao, D. R. Smith, Description and explanation of electromagnetic behaviors in artificial metamaterials based on effective medium theory, *Phys. Rev. E* 76, 026606 (2007).
139. C. L. Holloway, E. F. Kuester, J. A. Gordon, J. O'Hara, J. Booth, D. R. Smith, An overview of the theory and applications of metasurfaces, the two-dimensional equivalents of metamaterials, *IEEE Trans. Antennas Propag. Mag.* 54, 10–35 (2012).
140. T. J. Cui, M. Q. Qi, X. Wan, J. Zhao, Q. Cheng, Lights coding metamaterials, digital metamaterials and programmable metamaterials, *Light: Sci. Appl.* 3, e214 (2014).
141. P. Nayeri, F. Yang, A. Elsherbeni, Beam-scanning reflectarray antennas: A technical overview and state of the art, *IEEE Trans. Antennas Propag. Mag.* 57, 32–47 (2015).
142. P. Nayeri, F. Yang, A. Elsherbeni, Bifocal design and aperture phase optimizations of reflectarray antennas for wide-angle beam scanning performance, *IEEE Trans. Antennas Propag.* 61, 4588–4597 (2013).
143. T. J. Cui, S. Liu, L. Zhang, Information metamaterials and metasurfaces, *J. Mater. & Chem. C* 5, 3644–3668 (2017).
144. S. Liu, T. J. Cui, Concepts, working principles, and applications of coding and programmable metamaterials, *Adv. Opt. Mater.* 5(22), 1700624 (2017).

145. L. H. Gao, Q. Cheng, J. Yang, S. J. Ma, J. Zhao, S. Liu, H. B. Chen, Q. He, W. X. Jiang, H. F. Ma, Q. Y. Wen, L. J. Liang, B. B. Jin, W. W. Liu, L. Zhou, J. Q. Yao, P. H. Wu, T. J. Cui, Broadband diffusion of terahertz waves by multi-bit coding metasurfaces, *Light: Sci. Appl.* 2015, 4, e324 (2015).
146. L. J. Liang, M. Q. Qi, J. Yang, X. P. Shen, J. Q. Zhai, W. Z. Xu, B. B. Jin, W. W. Liu, Y. J. Feng, C. H. Zhang, H. Lu, H. T. Chen, L. Kang, W. W. Xu, J. Chen, T. J. Cui, P. H. Wu, S. G. Liu, Anomalous terahertz reflection and scattering by flexible and conformal coding metamaterials, *Adv. Opt. Mater.* 3, 1374–1380 (2015).
147. S. Liu, A. Noor, L. L. Du, L. Zhang, Q. Xu, K. L. T. Q. Wang, Z. Tian, W. X. Tang, J. G. Han, W. L. Zhang, X. Y. Zhou, Q. Cheng, T. J. Cui, Anomalous refraction and nondiffractive Bessel-beam generation of terahertz waves through transmission-type coding metasurfaces, *ACS Photon*. 3, 1968–1977 (2016).
148. Z. W. Wang, Q. Zhang, K. Zhang, G. K. Hu, Tunable digital metamaterial for broadband vibration isolation at low frequency, *Adv. Mater.* 28, 9857–9861 (2016).
149. B. Y. Xie, K. Tang, H. Cheng, Z. Y. Liu, S. Q. Chen, J. G. Tian, Coding acoustic metasurfaces, *Adv. Mater.* 29, 1603507 (2016).
150. B. Y. Xie, H. C. Tang, Z. Y. Liu, S. Q. Chen, J. G. Tian, Multiband asymmetric transmission of airborne sound by coded metasurfaces, *Phys. Rev. Appl.* 7, 024010 (2017).
151. S. Saadat, M. Adnan, H. Mosallaei, E. Afshari, Composite metamaterial and metasurface integrated with non-foster active circuit elements: A bandwidth-enhancement investigation, *IEEE Trans. Antennas Propag.* 61, 1210–1218 (2013).
152. M. Barbuto, A. Monti, F. Bilotti, A. Toscano, Design of a non-foster actively loaded SRR and application in metamaterial-inspired components, *IEEE Trans. Antennas Propag. Mag.* 61, 1219–1227 (2012).
153. S. Hrabar, I. Krois, I. Bonic, A. Kiricenko, Ultra-broadband simultaneous superluminal phase and group velocities in non-Foster epsilon-near-zero metamaterial, *Appl. Phys. Lett.* 102, 054108 (2013).
154. R. Elliott, Azimuthal surface waves on circular cylinders, *Trans. IRE Profession. Group Antennas Propagat.* 2, 71–81 (1954).
155. D. Sievenpiper, L. Zhang, R. F. J. Broas, N. Alexopolous, E. Yablonovitch, High-impedance electromagnetic surfaces with a forbidden frequency band, *IEEE Trans. Microwave Theory Techn.* 47, 2059–2074 (1999).

156. S. Liu, T. J. Cui, L. Zhang, Q. Xu, Q. Wang, X. Wan, J. Q. Gu, W. X. Tang, M. Q. Qi, J. G. Han, W. L. Zhang, X. Y. Zhou, Q. Cheng, Convolution operations on coding metasurface to reach flexible and continuous controls of terahertz beams., *Adv. Sci.* 3, 1600156 (2016).
157. S. Liu, T. J. Cui, Flexible controls of scattering clouds using coding metasurfaces, *Sci. Rep.* 6, 37545 (2016).
158. S. Liu, T. J. Cui, Flexible controls of terahertz waves using coding and programmable metasurfaces, *IEEE J. Sel. Top. Quantum Electron.* 23, 1–12 (2016).
159. M. Moccia, S. Liu, R. Y. Wu, G. Castaldi, A. Andreone, T. J. Cui, V. Galdi. Coding metasurfaces for diffuse scattering: Scaling laws, bounds, and sub-optimal design. *Adv. Opt. Mater.* 5, 1700455 (2017).
160. Y. R. Padooru, A. B. Yakovlev, P. Y. Chen, A. Alù, Analytical modeling of conformal mantle cloaks for cylindrical objects using sub-wavelength printed and slotted arrays, *J. Appl. Phys.* 112, 034907 (2012).
161. D. Rainwater, A. Kerkhoff, K. Melin, J. Soric, G. Moreno, A. Alù, Experimental verification of three-dimensional plasmonic cloaking in free-space, *New J. Phys.* 14, 013054 (2012).
162. K. Tang, C. Qiu, J. Lu, M. Ke, Z. Liu, Focusing and directional beaming effects of airborne sound through a planar lens with zigzag slits, *J. Appl. Phys.* 17, 024503 (2015).
163. Z. Yang, F. Gao, X. Shi, X. Lin, Z. Gao, Y. Chong, B. Zhang, Topological acoustics, *Phys. Rev. Lett.* 114, 114301(2015).
164. M. Xiao, W.-J. Chen, W.-Y. He, C. T. Chan, Synthetic gauge flux and Weyl points in acoustic systems, *Nat. Phys.* 11, 920 (2015).
165. A. B. Khanikaev, R. Fleury, S. H. Mousavi, A. Alù, Topologically robust sound propagation in an angular momentum-biased graphene-like resonator lattice, *Nat. Commun.* 6, 8260 (2015).
166. S. H. Mousavi, A. B. Khanikaev, Z. Wang, Topologically protected elastic waves in phononic metamaterials, *Nat. Commun.* 6, 8682 (2015).
167. Z. Liu, X. Zhang, Y. Mao, Y. Y. Zhu, Z. Yang, C. T. Chan, P. Sheng, Locally resonant sonic materials, *Science* 289, 1734–1736 (2000).
168. N. Fang, D. Xi, J. Xu, M. Ambati, W. Srituravanich, C. Sun, X. Zhang, Ultrasonic metamaterials with negative modulus, *Nat. Mater.* 5, 452–456 (2006).
169. R. S. Lakes, Advances in negative Poisson's ratio materials, *Adv. Mater.* 5, 293–296 (1993).
170. A. Bergamini, T. Delpero, L. de Simoni, L. di Lillo, M. Ruzzene, P. Ermanni, Phononic crystal with adaptive connectivity, *Adv. Mater.* 26, 1343–1347 (2014).

171. R. Zhu, Y. Y. Chen, M. V. Barnhart, G. K. Hu, C. T. Sun, G. L. Huang, Experimental study of an adaptive elastic metamaterial controlled by electric circuits, *Appl. Phys. Lett.* 108, 011905 (2016).
172. F. Casadei, T. Delpero, A. Bergamini, P. Ermanni, M. Ruzzene, Piezoelectric resonator arrays for tunable acoustic waveguides and metamaterials, *J. Appl. Phys.* 112, 064902 (2012).
173. S. Babaee, N. Viard, P. Wang, N. X. Fang, K. Bertoldi, Harnessing deformation to switch on and off the propagation of sound, *Adv. Mater.* 28, 1631–1635 (2016).
174. P. Wang, F. Casadei, S. Shan, J. C. Weaver, K. Bertoldi, Harnessing buckling to design tunable locally resonant acoustic metamaterials, *Phys. Rev. Lett.* 113, 014301 (2014).
175. Q. Zhang, D. Yan, K. Zhang, G. Hu, Pattern transformation of heat-shrinkable polymer by three-dimensional (3D) printing technique, *Sci. Rep.* 5, 8936 (2015).
176. Q. Zhang, K. Zhang, G. Hu, Smart three-dimensional lightweight structure triggered from a thin composite sheet via 3D printing technique, *Sci. Rep.* 6, 22431 (2016).
177. S. Liu, L. Zhang, Q. L. Yang, Q. Xu, Y. Yang, A. Noor, Q. Zhang, S. Iqbal, X. Wan, Z. Tian, W. X. Tang, Q. Cheng, J. G. Han, W. L. Zhang, T. J. Cui, Frequency-dependent dual-functional coding metasurfaces at terahertz frequencies, *Adv. Opt. Mater.* 4, 1965–1973 (2016).
178. L. X. Liu, X. Q. Zhang, M. Kenney, X. Q. Su, N. N. Xu, C. M. Ouyang, Y. Shi, J. G. Han, W. L. Zhang, S. Zhang, Broadband metasurfaces with simultaneous control of phase and amplitude. *Adv. Mater.* 26, 5031–5036 (2014).
179. X. Q. Zhang, Z. Tian, W. Yue, J. Gu, S. Zhang, J. Hanand, W., Zhang, Broadband terahertz wave deflection based on C-shape complex metamaterials with phase discontinuities. *Adv. Mater.* 2013, 25, 4567–4572.
180. S. Liu, H. C. Zhang, L. Zhang, Q. Xu, Q. L. Yang, J. Q. Gu, H. F. Ma, W. X. Jiang, X. Y. Zhou, J. G. Han, W. L. Zhang, Q. Cheng, T. J. Cui, Full-state controls of terahertz waves using tensor coding metasurfaces, *ACS Appl. Mater. & Interfaces* 9, 21503–21514, (2017).
181. S. L. Sun, Q. He, S. Y. Xiao, Q. Xu, X. Li, Lei Zhou, Gradient-index meta-surfaces as a bridge linking propagating waves and surface waves, *Nat. Mater.* 11, 426–431 (2011).
182. H. H. Yang, F. Yang, S. H. Xu, Y. L. Mao, M. K. Li, X. Y. Cao, J. Gao, A 1-bit 10 × 10 reconfigurable reflectarray antenna: design, optimization, and experiment, *IEEE Trans. Antennas Propag.* 64, 2246–2254 (2016).

183. X. Wan, M. Q. Qi, T. Q. Chen, T. J. Cui, Field-programmable beam reconfiguring based on digitally-controlled coding metasurface, *Sci. Rep*. 6, 20663 (2016).
184. H. H. Yang, X. Y. Cao, F. Yang, J. Gao, S. H. Xu, M. K. Li, X. B. Chen, Y. Zhao, Y. J. Zheng, S. J. Li, A programmable metasurface with dynamic polarization, scattering and focusing control, *Sci. Rep*. 6, 35692 (2016).
185. H. Kamoda, T. Iwasaki, J. Tsumochi, T. Kuki, O. Hashimoto, 60-GHz electronically reconfigurable large reflectarray using single-bit phase shifters, *IEEE Trans. Antennas Propag*. 59, 2524–2531 (2011).
186. C. E. Shannon, A mathematical theory of communication, *ACM SIGMOBILE Mobile Computing and Communications Review* 5, 3–55 (2001).
187. T. J. Cui, S. Liu, L. L. Li, Information entropy of coding metasurface, *Light: Sci. Appl*. 5, e16172 (2016).
188. R. Y. Wu, C. B. Shi, S. Liu, W. Wu, T. J. Cui. Addition theorem for digital coding metamaterials, *Adv. Opt. Mater*. 1701236 (2018).
189. M. F. Duarte, M. A. Davenport, D. Takhar, J. N. Laska, T. Sun, K. F. Kelly, R. G. Baraniuk, Single-pixel imaging via compressive sampling, *IEEE Signal Process. Mag*. 25, 83–91 (2008).
190. C. M. Watts, D. Shrekenhamer, J. Montoya, G. Lipworth, J. Hunt, T. Sleasman, S. Krishna, D. R. Smith, W. J. Padilla, Terahertz compressive imaging with metamaterial spatial light modulators. *Nat. Photonics*. 8, 605–609 (2014).
191. G. Lipworth, A. Mrozack, J. Hunt, D. L. Marks, T. Driscoll, D. Brady, D. R. Smith, Metamaterial apertures for coherent computational imaging on the physical layer. *J. Opt. Soc. Am. A* 30, 1603–1612 (2013).
192. J. Hunt, J. Gollub, T. Driscoll, et al. Metamaterial microwave holographic imaging system, *J. Opt. Soc. Am*. A 31, 2109 (2014).
193. C. M. Watts, X. Liu, W. J. Padilla, Metamaterial electromagnetic wave absorbers, *Adv. Opt. Mater*. 24, 98–120 (2012).
194. B. Sensale-Rodriguez, S. Rafique, R. Yan, M. Zhu, V. Protasenko, D. Jena, L. Liu, H. L. G. Xing, Terahertz imaging employing graphene modulator arrays, *Opt. Express*, 21(2),2324–2330 (2013).
195. Y. B. Li, L. L. Li, B. B. Xu, W. Wu, R. Y. Wu, X. Wan, Q. Cheng, T. J. Cui, Transmission-type 2-bit programmable metasurface for single-sensor and single-frequency microwave imaging, *Sci. Rep*. 6, 23731 (2015).
196. L. Li, M. Hurtado, F. Xu, B. C. Zhang, T. Jin, T. J. Cui, M. N. Stevanovic, A. Nehorai, A survey on the low-dimensional-model-based electromagnetic

imaging, *Foundations and Trends in Signal Processing* 12(2),107–199 (2018).

197. B. Walther, C. Helgert, C. Rockstuhl, et al. Spatial and spectral light shaping with metamaterials, *Adv. Mater.* 24, 6300–6304 (2012).
198. B. Gholipour, J. Zhang, K. F. MacDonald, D. W. Hewak, N. I., Zheludev, An all-optical, non-volatile, bidirectional, phase-change meta-switch, *Adv. Mater.* 25, 3050–3054 (2013).
199. Q. Wang, E. T. F. Rogers, B. Gholipour, C. M. Wang, G. H. Yuan, J. H. Teng, N. I. Zheludev, Optically reconfigurable metasurfaces and photonic devices based on phase change materials, *Nat. Photon.* 10, 60–65 (2016).
200. G. Kaplan, K. Aydin, J. Scheuer, Dynamically controlled plasmonic nano-antenna phased array utilizing vanadium dioxide, *Opt. Mater. Express* 5, 2513 (2015).
201. M. J. Dicken, K. Aydin, I. M. Pryce, L. A. Sweatlock, E. M. Boyd, S. Walavalkar, J. Ma, H. A. Atwater, Frequency tunable near-infrared metamaterials based on VO2 phase transition, *Opt. Express* 17, 18330 (2009).
202. H. Tao, A. C. Strikwerda, K. Fan, W. J. Padilla, X. Zhang, R. D. Averitt, Reconfigurable terahertz metamaterials. *Phys. Rev. Lett.* 103, 147401 (2009).
203. L. Ju, B. S. Geng, J. Horng, C. Girit, M. Martin, Z. Hao, H. A. Bechtel, X. G. Liang, A. Zettl, T. R. Shen, F. Wang, Graphene plasmonics for tunable terahertz metamaterials. *Nat. Nanotechnol.* 6, 630–634 (2011).
204. Y.-W. Huang, H. W. H. Lee, R. Sokhoyan, R. A. Pala, K. Thyagarajan, S. Han, D. P. Tsai, H. A. Atwater, Gate-tunable conducting oxide metasurfaces. *Nano Lett.* 16, 5319–5325 (2016).
205. L. Li, T. J. Cui, W. Ji, S. Liu, J. Ding, X. Wan, Y. B. Li, M. Jiang, C.-W. Qiu, S. Zhang, Electromagnetic reprogrammable coding metasurface holograms, *Nat. Commun.* 8, 197 (2017).
206. R. W. Gerchberg, W. O. Saxton, A practical algorithm for the determination of the phase from image and diffraction plane pictures. *Optik* 35, 227–246 (1972).
207. J. Zhao, X. Yang, J. Y. Dai, Q. Cheng, X. Li, N. H. Qi, J. C. Ke, G. D. Bai, S. Liu, S. Jin, A. Alu, T. J. Cui, Programmable time-domain digital-coding metasurface for non-linear harmonic manipulation and new wireless communication systems, *National Science Review* 6(2), 231–238 (2019).
208. A. Silva, F. Monticone, G. Castaldi, V. G. Baldi, A. Alù, N. Engheta, Performing mathematical operations with metamaterials, *Science*, 343, 160–163 (2014).
209. D. Shrekenhamer, W. C. Chen, W. J. Padilla, Liquid crystal tunable metamaterial absorber, *Phys. Rev. Lett.* 110, 177403 (2013).

210. X. Liu, W. J. Padilla, Dynamic manipulation of infrared radiation with MEMS metamaterials, *Adv. Opt. Mater.* 1, 559–562 (2013).
211. S. F. Shi, B. Zeng, H. L. Han, X. Hong, H. Z. Tsai, A. Zettl, M. F. Crommie, F. Wang, Optimizing broadband terahertz modulation with hybrid graphene/metasurface structures, *Nano Lett.* 15, 372 (2014).
212. X. Liu, J. Q. Gu, R. Singh, Y. F. Ma, J. Zhu, Z. Tian, M. X. He, J. G. Han, W. L. Zhang, Electromagnetically induced transparency in terahertz plasmonic metamaterials via dual excitation pathways of the dark mode, *Appl. Phys. Lett.* 100, 131101 (2012).
213. J. Gu, R. Singh, X. J. Liu, X. Q. Zhang, Y. F. Ma, S. Zhang, S. A. Maier, Z. Tian, A. K. Azad, H. T. Chen, A. J. Taylor, J. G. Han, W. L. Zhang, Active control of electromagnetically induced transparency analog in terahertz metamaterials, *Nat. Commun.* 3, 1151 (2012).
214. H. T. Chen, J. F. O'Hara, A. K. Azad, A. J. Taylor, R. D. Averitt, D. B. Shrekenhamer, W. J. Padilla, Experimental demonstration of frequency-agile terahertz metamaterials, *Nat. Photon.* 2, 295–298 (2008).
215. H. T. Chen, H. Yang, R. Singh, J. F. O'Hara, A. K. Azad, S. A. Trugman, Q. X. Jia, A. J. Taylor, Tuning the resonance in high-temperature superconducting terahertz metamaterials, *Phys. Rev. Lett.* 105, 247402 (2010).
216. W. M. Zhu, Q. H. Song, L. B. Yan, W. Zhang, P. C. Wu. L. K. Chin, H. Cai, D. P. Tsai, Z. X. Shen, T. W. Deng, S. K. Ting, Y. D. Gu, G. Q. Lo, D. L. Kwong, Z. C. Yang, R. Huang, A. Q. Liu, N. Zheludev, A flat lens with tuneable phase gradient by using random access reconfigurable metamaterial, *Adv. Mater.* 27, 4739–4743 (2015).
217. P. C. Wu, W. M. Zhu, Z. X. Shen, P. H. J. Chong, W. Ser. D. P. Tsai, A. Q. Liu, Broadband wide-angle multifunctional polarization converter via liquid-metal-based metasurface, *Adv. Opt. Mater.* 5, 1600938 (2017).
218. L. B. Yan, W. M. Zhu, P. C. Wu, H. Cai, Y. D. Gu, L. K. Chin, Z. X. Shen, P. H. J. Chong, Z. C. Yang, W. Ser, D. P. Tsai, A. Q. Liu, Adaptable metasurface for dynamic anomalous reflection, *Appl. Phys. Lett.* 110, 201904 (2017).
219. X. Yang, S. H. Xu, F. Yang, M. K. Li, Y. Q. Hou, S. D. Jiang, L. Liu, A broadband high-efficiency reconfigurable reflectarray antenna using mechanically rotational elements, *IEEE Trans. Antennas Propag. Mag.* 65, 3959–3966 (2017).
220. V. F. Fusco, Mechanical beam scanning reflectarray, *IEEE Trans. Antennas Propag.* 53, 3842–3844 (2005).
221. B. Subbarao, V. Srinivasan, V. F. Fusco, R. Cahill, Element suitability for circularly polarised phase agile reflectarray applications, *IEE Proc.-Microw., Antennas Propag.* 151, 287–292 (2004).

222. V. Srinivasan, V. F. Fusco, Circularly polarised mechanically steerable reflectarray, *IEE Proc.-Microw. Antennas Propag.* 152, 511–514 (2005).
223. J. Y. Ou, E. Plum, J. F. Zhang, N. I. Zheludev, An electromechanically reconfigurable plasmonic metamaterial operating in the near-infrared. *Nat. Nanotech.* 8, 252 (2013).
224. L. Zhang, X. Q. Chen, S. Liu, Q. Zhang, J. Zhao, J. Y. Dai, G. D. Bai, X. Wan, Q. Cheng, G. Castaldi, V. Galdi, T. J. Cui, Space-time-coding digital metasurfaces, *Nat. Commun.* 9, 4334 (2018).
225. L. Zhang, Z. X. Wang, R. W. Shao, J. L. Shen, X. Q. Chen, X. Wan, Q. Cheng, T. J. Cui, Dynamically realizing arbitrary multi-bit programmable phases using a 2-bit time-domain coding metasurface, *IEEE Transactions on Antennas and Propagation* 67, DOI 10.1109/TAP.2019.2955219 (2019).
226. L. Zhang, X. Q. Chen, R. W. Shao, J. Y. Dai, Q. Cheng, G. Castaldi, V. Galdi, T. J. Cui, Breaking reciprocity with space-time-coding digital metasurfaces, *Adv. Mater.* 31, 1904069 (2019).
227. L. Bao, Q. Ma, G. D. Bai, H. B. Jing, R. Y. Wu, C. Yang, J. Wu, X. Fu, T. J. Cui, Design of digital coding metasurfaces with independent controls of phase and amplitude responses, *Appl. Phys. Lett.* 113, 063502 (2018).
228. J. Luo, Q. Ma, H. B. Jing, G. D. Bai, R. Y. Wu, L. Bao, T. J. Cui, 2-bit amplitude-modulated coding metasurfaces based on indium tin oxide films, *J. Appl. Phys.* 126, 113102 (2019).
229. R. Y. Wu, L. Zhang, L. Bao, L. W. Wu, Q. Ma, G. D. Bai, H. T. Wu, T. J. Cui, Digital metasurface with phase code and reflection-transmission amplitude code for flexible full-space electromagnetic manipulations, *Adv. Opt. Mater.* 7, 1801429 (2019).
230. L. Bao, R. Y. Wu, X. Fu, Q. Ma, G. D. Bai, T. J. Cui, Multi-beam forming and controls by metasurface with phase and amplitude modulations, *IEEE Transactions on Antennas and Propagation* 67(10),6680–6685 (2019).
231. L. Chen, Q. Ma, H. B. Jing, H. Y. Cui, Y. Liu, T. J. Cui, Spatial-energy digital coding metasurface based on active amplifier, *Phys. Rev. Appl.* 11, 054051 (2019).
232. Z. Luo, M. Z. Chen, Z. X. Wang, L. Zhou, Q. Wang, Y. B. Li, Q. Cheng, H. F. Ma, T. J. Cui, Digital nonlinear metasurface with highly customizable nonreciprocity, *Adv. Funct. Mater.* 29, 1906635 (2019).
233. Q. Ma, L. Chen, H. B. Jing, Q. R. Hong, H. Y. Cui, Y. Liu, L. Li, T. J. Cui, Controllable and programmable nonreciprocity based on detachable digital coding metasurface *Adv. Opt. Mater.* 7, 1901285 (2019).
234. Z. Luo, Q. Wang, X. G. Zhang, J. W. Wu, J. Y. Dai, L. Zhang, H. T. Wu, H. C. Zhang, H. F. Ma, Q. Cheng, T. J. Cui, Intensity-dependent

metasurface with digitally-reconfigurable distribution of nonlinearity, *Adv. Opt. Mater.* 7, 1900792 (2019).

235. L. Li, H. Ruan, C. Liu, Y. Li, Y. Shuang, A. Alù, C.-W. Qiu, T. J. Cui, Machine-learning reprogrammable metasurface imager, *Nat. Commun.* 10, 1082 (2019).
236. L. Li, Y. Shuang, Q. Ma, H. Li, H. Zhao, M. Wei, C. Liu, C. Hao, C. W. Qiu, T. J. Cui, Intelligent metasurface imager and recognizer, *Light: Sci. Appl.* 8, 97 (2019).
237. H. Y. Li, H. T. Zhao, M. L. Wei, et al. Intelligent electromagnetic sensing with learnable data acquisition and processing, *Patterns* 1, 100006, (2020).
238. Q. Ma, G. D. Bai, H. B. Jing, C. Yang, L. Li, T. J. Cui, Smart metasurface with self-adaptively reprogrammable functions, *Light: Sci. Appl.* 8, 98 (2019).
239. T. J. Cui, S. Liu, G. D. Bai, Q. Ma, Direct transmission of digital message via programmable coding metasurface. *Research* 2584509 (2019).

Cambridge Elements

Emerging Theories and Technologies in Metamaterials

Tie Jun Cui
Southeast University, China
Tie Jun Cui is Cheung-Kong Professor and Chief Professor at Southeast University, China, and a Fellow of the IEEE. He has made significant contributions to the area of effective-medium metamaterials and spoof surface plasmon polaritons at microwave frequencies, both in new-physics verification and engineering applications. He has recently proposed digital coding, field-programmable, and information metamaterials, which extend the concept of metamaterial.

John B. Pendry
Imperial College London
Sir John Pendry is Chair in Theoretical Solid State Physics at Imperial College London, and a Fellow of the Royal Society, the Institute of Physics and the Optical Society of America. Among his many achievements are the proposal of the concepts of an 'invisibility cloak' and the invention of the transformation optics technique for the control of electromagnetic fields.

About the Series

Bringing together viewpoints of leading scientists and engineers, this new series provides systematic coverage of new and emerging topics in metamaterials. It covers the theory, characterisation, design and fabrication of metamaterials in a wide expanse of areas also showcases the very latest experimental techniques and applications.
This series is perfect for graduate students, researchers, and professionals with a background in physics and electrical engineering.

Cambridge Elements

Emerging Theories and Technologies in Metamaterials

Elements in the Series

Spoof Surface Plasmon Metamaterials
Paloma Arroyo Huidobro, Antonio I. Fernández-Domíguez, John B. Pendry, Luis Martín-Moreno, and Francisco J. Garcia-Vidal

Metamaterials and Negative Refraction
Rujiang Li, Zuojia Wang, and Hongsheng Chen

Information Metamaterials
Tie Jun Cui and John B. Pendry

A full series listing is available at: www.cambridge.org/EMMA

Printed in the United States
By Bookmasters